Praise for *The Earth Chronicles* series

"Exciting . . . credible . . . most provocative and compelling."
—*Library Journal*

"A dazzling performance . . . Sitchin is a zealous investigator."
—*Kirkus Reviews*

"Several factors make Sitchin's well-referenced works outstandingly different from all others that present this central theme. For one, his linguistic skills, which include not only several modern languages that make it possible for him to consult other scholars' works in their original tongues, but the ancient Sumerian, Egyptian, Hebrew, and other languages of antiquity as well.

"The devotion of thirty years to academic search and personal investigation before publishing resulted in unusual thoroughness, perspective, and modifications where need arose. The author's pursuit of the earliest available texts and artifacts also made possible the wealth of photos and line drawings made for his books from tablets, monuments, murals, pottery, seals, etc. Used generously throughout, they provide vital visual evidence. . . . While the author does not pretend to solve all the puzzles that have kept intensive researchers baffled for well over one hundred years, he has provided some new clues."

—Rosemary Decker, historian and researcher

ZECHARIA SITCHIN

When Time Began

The Fifth
Book of
The Earth Chronicles

Bear & Company
Rochester, Vermont

Bear & Company
One Park Street
Rochester, Vermont 05767
www.InnerTraditions.com

Bear & Company is a division of Inner Traditions International

Copyright © 1993, 1994 by Zecharia Sitchin

Library of Congress Cataloging-in-Publication Data

Sitchin, Zecharia.
    When time began / Zecharia Sitchin.
        p. cm. — (The fifth book of the Earth chronicles)
    Reprint. Originally published: New York : Avon Books, 1993.
    Includes bibliographical references and index.
    ISBN 978-187918116-8
    1. Civilization, ancient—Extraterrestrial influences. 2. Stonehenge
(England). I. Title. II. Series: Sitchin, Zecharia. Earth chronicles ; 5.
CB 156.S5931994
930—dc20   94-2823
                                                           CIP

Printed and bound in India

10 9 8 7 6 5 4 3

CONTENTS

CONTENTS

FOREWORD

Since the earliest times, Earthlings have lifted their eyes unto the heavens. Awed as well as fascinated, Earthlings learned the Ways of Heaven: the positions of the stars, the cycles of Moon and Sun, the turning of an inclined Earth. How did it all begin, how will it end—and what will happen in between?

Heaven and Earth meet on the horizon. For millennia Earthlings have watched the stars of the night give way to the rays of the Sun at that meeting place, and chose as a point of reference the moment when daytime and nighttime are equal, the day of the Equinox. Man, aided by the calendar, has counted Earthly Time from that point on.

To identify the starry heavens, the skies were divided into twelve parts, the twelve houses of the zodiac. But as the millennia rolled on, the "fixed stars" seemed not to be fixed at all, and the Day of the Equinox, the day of the New Year, appeared to shift from one zodiacal house to another; and to Earthly Time was added Celestial Time— the start of a new era, a New Age.

As we stand at the threshold of a New Age, when sunrise on the day of the spring equinox will occur in the zodiacal house of Aquarius rather than, as in the past 2,000 years, in the zodiacal house of Pisces, many wonder what the change might portend: good or evil, a new beginning or an end—or no change at all?

To understand the future we should examine the past; because since Mankind began to count Earthly Time, it has already experienced the measure of Celestial Time—the arrival of New Ages. What preceded and followed one such New Age holds great lessons for our own present station in the course of Time.

1

THE CYCLES OF TIME

It is said that Augustine of Hippo, the bishop in Roman Carthage (A.D. 354–430), the greatest thinker of the Christian Church in its early centuries, who fused the religion of the New Testament with the Platonistic tradition of Greek philosophy, was asked, "What is time?" His answer was, "If no one asks me, I know what it is; if I wish to explain what it is to him who asks me, I do not know."

Time is essential to Earth and all that is upon it, and to each one of us as individuals; for, as we know from our own experience and observations, what separates us from the moment we are born and the moment when we cease to live is TIME.

Though we know not what Time is, we have found ways to measure it. We count our lifetimes in *years,* which—come to think of it—is another way of saying "orbits," for that is what a "year" on Earth is: the time it takes Earth, our planet, to complete one orbit around our star, the Sun. We do not know what time is, but the way we measure it makes us wonder: would we live longer, would our life cycle be different, were we to live on another planet whose "year" is longer? Would we be "immortal" if we were to be upon a "Planet of millions of years"—as, in fact, the Egyptian pharaohs believed that they would be, in an eternal Afterlife, once they joined the gods on that "Planet of millions of years"?

Indeed, are there other planets "out there," and, even more so, planets on which life as we know it could have evolved—or is our planetary system unique, and life on Earth unique, and we, humankind, are all alone—or did the pharaohs know what they were speaking of in their Pyramid Texts?

2

"Look up skyward and count the stars," Yahweh told Abraham as He made the covenant with him. Man has looked skyward from time immemorial, and has been wondering whether there are others like him out there, upon other earths. Logic, and mathematical probability, dictate a Yes answer; but it was only in 1991 that astronomers, *for the first time,* it was stressed, actually found other planets orbiting other suns elsewhere in the universe.

The first discovery, in July 1991, turned out not to have been entirely correct. It was an announcement by a team of British astronomers that, based on observations over a five-year period, they concluded that a rapidly spinning star identified as Pulsar 1829–10 has a "planet-sized companion" about ten times the size of Earth. Pulsars are assumed to be the extraordinarily dense cores of stars that have collapsed for one reason or another. Spinning madly, they emit pulses of radio energy in regular bursts, many times per second. Such pulses can be monitored by radio telescopes; by detecting a cyclic fluctuation, the astronomers surmised that a planet that orbits Pulsar 1829–10 once every six months can cause and explain the fluctuation.

As it turned out, the British astronomers admitted several months later that their calculations were imprecise and, therefore, they could not stand by their conclusion that the pulsar, some 30,000 light-years away, had a planetary satellite. By then, however, an American team had made a similar discovery pertaining to a much closer pulsar, identified as PSR 1257 + 12—a collapsed sun only 1,300 light-years away from us. It exploded, astronomers estimated, about a mere billion years ago; and it definitely has two, and perhaps three, orbiting planets. The two certain ones were orbiting their sun at about the same distance as Mercury does our Sun; the possible third planet orbits its sun at about the same distance as Earth does our Sun.

"The discovery stirred speculation that planetary systems not only were fairly common but also could occur under diverse circumstances," wrote John Noble Wilford in *The New York Times* of January 9, 1992; "scientists said it was most unlikely that planets orbiting pulsars could be hospitable to life; but the findings encouraged astronomers, who this fall will begin a systematic survey of the heavens for signs of intelligent extraterrestrial life."

Were, then, the pharaohs right?

Long before the pharaohs and the Pyramid Texts, an ancient civilization—Man's first known one—possessed an advanced cosmogony. Six thousand years ago, in ancient Sumer, what astronomers have discovered in the 1990s was already known; not only the true nature and composition of our Solar System (including the farthest out planets), but also the notion that there are other solar systems in the universe, that their stars ("suns") can collapse or explode, that their planets can be thrown off course—that Life, indeed, can thus be carried from one star system to another. It was a detailed cosmogony, spelled out in writing.

One long text, written on seven tablets, has reached us primarily in its later Babylonian version. Called the *Epic of Creation* and known by its opening words *Enuma elish,* it was publicly read during the New Year festival that started on the first day of the month Nissan, coinciding with the first day of spring.

Outlining the process by which our own Solar System came into being, the long text described how the Sun ("Apsu") and its messenger Mercury ("Mummu") were first joined by an olden planet called Tiamat; how a pair of planets Venus and Mars—("Lahamu" and "Lahmu") then coalesced between the Sun and Tiamat, followed by two pairs beyond Tiamat—Jupiter and Saturn ("Kishar" and "Anshar") and Uranus and Neptune ("Anu" and "Nudim-mud"), the latter two being planets unknown to modern astronomers until 1781 and 1846 respectively—yet known, and described, by the Sumerians millennia earlier. As those newly-created "celestial gods" tugged and pulled at each other, some of them sprouted satellites—moonlets. Tiamat, in the midst of that unstable planetary family, sprouted eleven satellites; one of them, "Kingu," grew so much in size that it began to assume the aspects of a "celestial god," a planet, on its own. Modern astronomers were totally ignorant of the possibility that a planet could have many moons until Galileo discovered the four largest moons of Jupiter in 1609, with the aid of a telescope; but the Sumerians were aware of the phenomenon millennia earlier.

Into that unstable solar system, according to the millennia-old *Epic of Creation*, there appeared an invader from outer space—another planet; a planet not born into the family of Apsu, but one that had belonged to some other star's family and that was thrust off to wander in space. Millennia before modern astronomy learned of pulsars and collapsing stars, the Sumerian cosmogony had already envisioned other planetary systems and collapsing or exploding stars that threw off their planets. And so, *Enuma elish* related, one such cast-off planet, reaching the outskirts of our own Solar System, began to be drawn into its midst (Fig. 1).

As it passed by the outer planets, it caused changes that account for many of the enigmas that still baffle modern astronomy—such as the cause for Uranus's tilt on its side, the retrograde orbit of Neptune's largest moon, Triton, or what pulled Pluto from its place as a moonlet to become a planet with an odd orbit. The more the invader was drawn into the Solar System's center, the more was it forced onto a collision course with Tiamat, resulting in the "Celestial Battle." In the series of collisions, with the invader's sat-

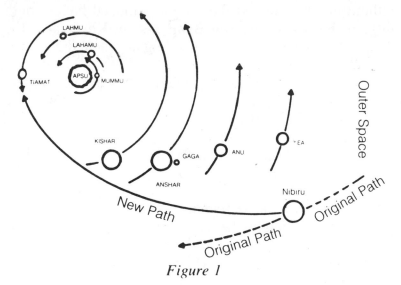

Figure 1

ellites repeatedly smashing into Tiamat, the olden planet split in two. One half of it was smashed into bits and pieces to become the Asteroid Belt (between Mars and Jupiter) and various comets; the other half, wounded but intact, was thrust into a new orbit to become the planet we call Earth ("Ki" in Sumerian); shunted with it was Tiamat's largest satellite, to become Earth's Moon. The invader itself was caught into permanent orbit around the Sun, to become our Solar System's twelfth member (Sun, Moon, and ten planets). The Sumerians called it *Nibiru*—"Planet of the Crossing." The Babylonians renamed it *Marduk* in honor of their national god. It was during the Celestial Battle, the ancient epic asserted, that the "seed of life," brought by Nibiru from elsewhere, was passed to Earth.

Philosophers and scientists, contemplating the universe and offering modern cosmogonies, invariably end up discussing Time. Is Time a dimension in itself, or perhaps the only true dimension in the universe? Does Time only flow forward, or can it flow backward? Is the present part of the past or the beginning of the future? And, not least of all, did Time have a beginning? For, if so, will it have an end? If the universe has existed forever, without a beginning and thus without an end, is Time too without a beginning and without an end—or did the universe indeed have a beginning, perhaps with the Big Bang assumed by many astrophysicists, in which case Time began when the universe began?

Those who conceived the amazingly accurate Sumerian cosmogony also believed in a Beginning (and thus, inexorably, in an End). It is clear that they conceived of Time as a measure, the pacesetter from, and the marker of, a beginning in a celestial saga; for the very first word of the ancient Epic of Creation, *Enuma*, means *When:*

Enuma elish la nabu shamamu
   When in the heights heaven had not been named
Shaplitu ammatum shuma la zakrat
   And below, firm ground (Earth) had not been called

It must have taken great scientific minds to conceive of a primordial phase when "naught existed but primordial Apsu, their begetter; Mummu, and Tiamat"—when Earth had not yet come into being; and to realize that for Earth and all upon it the "big bang" was not when the universe or even the Solar System was created, but the event of the Celestial Battle. It was then, at that moment, that Time began for Earth—the moment when, separated from the half of Tiamat that became the Asteroid Belt ("heaven"), Earth was shunted to its own new orbit and could start counting the years, the months, the days, the nights—to measure Time.

This scientific view, central to ancient cosmogony, religion, and mathematics, was expressed in many other Sumerian texts besides the *Epic of Creation*. A text treated by scholars as the "myth" of "Enki and the world order," but which is literally the autobiographical tale by Enki, the Sumerian god of science, describes the moment when—*When*—Time began to tick for Earth:

> *In the days of yore,*
> *when heaven was separated from Earth,*
> *In the nights of yore,*
> *when heaven was separated from Earth . . .*

Another text, in words often repeated on Sumerian clay tablets, conveyed the notion of Beginning by listing the many aspects of evolution and civilization that had not yet come into being before the crucial event. Before then, the text asserted, "the name of Man had not yet been called" and "needful things had not yet been brought into being." All those developments started to take place only "after heaven had been moved away from Earth, after Earth had been separated from heaven."

It is not surprising that the same notions of Time's beginnings also ruled Egyptian beliefs, whose development was subsequent to those of the Sumerians. We read in the Pyramid Texts (para. 1466) the following description of the Beginning of Things:

When heaven had not yet come into existence,
When men had not yet come into existence,
When gods had not yet been born,
When death had not yet come into existence . . .

This knowledge, universal in antiquity and stemming from the Sumerian cosmogony, was echoed in the very first verse of Genesis, the first book of the Hebrew Bible:

In the beginning
Elohim *created the heaven and the earth.*
And the earth was without form and void
and darkness was upon the face of Tehom,
and the wind of the Lord swept over its waters.

It is now well established that this biblical tale of creation was based on Mesopotamian texts such as *Enuma elish,* with *Tehom* meaning Tiamat, the "wind" meaning "satellites," in Sumerian, and "heaven," described as the "hammered out bracelet," the Asteroid Belt. The Bible, however, is clearer regarding the moment of the Beginning as far as Earth was concerned; the biblical version picks up the Mesopotamian cosmogony only from the point of the separation of the Earth from the *Shama'im,* the Hammered Bracelet, as a result of the breakup of Tiamat.

For Earth, Time began with the Celestial Battle.

The Mesopotamian tale of creation begins with the formation of our Solar System and the appearance of Nibiru/Marduk at a time when the planetary orbits were not yet fixed and stable. It ends by attributing to Nibiru/Marduk the current shape of our Solar System, whereby each planet ("celestial god") received its assigned place ("station"), orbital path ("destiny"), and rotation, even its moons. Indeed, as a large planet that encompasses by its orbit all the other planets, one who "crosses the heavens and surveys the regions," it was considered to have become the one that stabilized the Solar System:

He established the station of Nibiru,
to determine their heavenly bands,
that none might transgress or fall short . . .

He established for the planets their
sacred heavens,
He keeps hold on their ways,
determines their courses.

Thus, states *Enuma elish* (Tablet V, line 65), "He created the Heaven and the Earth"—the very same words used in the Book of Genesis.

The Celestial Battle eliminated Tiamat as a member of the old Solar System, thrust half of it into a new orbit to become Planet Earth, retained the Moon as a vital component of the new Solar System, detached Pluto into an independent orbit, and added Nibiru as the twelfth member of the New Order in our heavens. For Earth and its inhabitants, those were to become the elements that determined Time.

To this day, the key role that the number twelve played in Sumerian science and daily life (in line with the twelve-member Solar System) has accompanied us throughout the millennia. They divided the "day" (from sunset to sunset) into twelve "double-hours," retained into modern times in the twelve-hour clock and the twenty-four-hour day. The twelve months in the year are still with us, as are the twelve houses of the zodiac. This celestial number had many other expressions, as in the twelve tribes of Israel or the twelve apostles of Jesus.

The Sumerian mathematical system is called sexagesimal, i.e. "based on sixty" rather than on 100 as in the metric system (in which one meter is equal to 100 centimeters). Among the advantages of the sexagesimal system was its divisibility into twelve. The sexagesimal system progressed by alternately multiplying six and ten: starting with six, multiplying six by ten ($6 \times 10 = 60$), then by six to obtain 360—the number applied by the Sumerians to the circle and still used both in geometry and astronomy. That, in turn, was multiplied by ten, to obtain the *sar* ("ruler, lord"), the number 3,600, which was written by inscribing a great circle, and so on.

The *sar,* 3,600 Earth-years, was the orbital period of Nibiru around the Sun; for anyone on Nibiru, it was just one Nibiru-year. According to the Sumerians, there were indeed others, intelligent beings, on Nibiru, evolving there well ahead of hominids on Earth. The Sumerians called them *Anunnaki,* literally meaning "Those who from Heaven to Earth came." Sumerian texts repeatedly asserted that the Anunnaki had come to Earth from Nibiru in great antiquity; and that when they had come here, they counted time not in Earth terms but in terms of Nibiru's orbit. The unit of that Divine Time, a year of the gods, was the *sar.*

Texts known as the Sumerian King Lists, which describe the first settlements of the Anunnaki on Earth, list the governorships of the first ten Anunnaki leaders before the Deluge in *sars,* the 3,600 Earth-year cycles. From the first landing to the Deluge, according to those texts, 120 *sars* had passed: Nibiru orbited the Sun one hundred and twenty times, which equals 432,000 Earth-years. It was on the one hundred twentieth orbit that the gravitational pull of Nibiru was such that it caused the ice sheet that accumulated over Antarctica to slip off into the southern oceans, creating the immense tidal wave that engulfed the Earth—the great flood or Deluge, recorded in the Bible from much earlier and much more detailed Sumerian sources.

Legends and ancient lore gave this number, 432,000, cyclical significance beyond the land then called Sumer. In *Hamlet's Mill,* Giorgio de Santillana and Hertha von Dechend, searching for "a point where myth and science join," concluded that "432,000 was a number of significance from old." Among the examples cited by them was the Teutonic and Norse tale of the Valhalla, the mythic abode of the slain warriors who, on the Day of Judgment, will march out of the Valhalla's gates to fight at the side of the god Odin or Woden against the giants. They would exit through the Valhalla's 540 doors; eight hundred warriors would march out of each. The total number of warrior-heroes, Santillana and von Dechend pointed out, was thus 432,000. "This number," they continued, "must have had a very ancient meaning, for it is also the number of syllables in the *Rigveda*," the "Sacred Book of Verses" in the Sanskrit language, in which have been recorded the Indo-European tales of gods and heroes. Four hundred thirty-two thousand, the two authors wrote, "goes back to the basic figure 10,800, the number of stanzas in the *Rigveda*, with 40 syllables to a stanza" (10,800 × 40 = 432,000).

Hindu traditions clearly associated the number 432,000 with the *yugas* or Ages that Earth and Mankind had experienced. Each *caturyuga* ("great yuga") was divided into four yugas or Ages whose diminishing lengths were expressions of 432,000: first the Fourfold Age (4 × 432,000 = 1,728,000 years) which was the Golden Age, then the Threefold Age of Knowledge (3 × 432,000 = 1,296,000 years), followed by the Double or Twofold Age of Sacrifice (2 × 432,000 = 864,000 years); and finally our present era, the Age of Discord which will last a mere 432,000 years. All in all these Hindu traditions envision ten eons, paralleling the ten Sumerian rulers of the pre-Diluvial era but expanding the overall time span to 4,320,000 years.

Further expanded, such astronomical numbers based on 432,000 were applied in Hindu religion and traditions to the *kalpa*, the "Day" of the Lord Brahma. It was defined as an eon comprising twelve million *devas* ("Divine Years"). Each Divine Year in turn equaled 360 Earth-years. Therefore, a "Day of the Lord Brahma" equaled 4,320,000,000 Earth-years—a time span very much like modern estimates of the age of our Solar System—arrived at by multiplications of 360 and 12.

4,320,000,000 is, however, a thousandfold great yugas— a fact brought out in the eleventh century by the Arab mathematician Abu Rayhan al-Biruni, who explained that the *kalpa* consisted of 1,000 cycles of caturyugas. One could thus paraphrase the mathematics of the Hindu celestial calendar by stating that in the eyes of the Lord Brahma, a thousand cycles were but a single day. This brings to mind the enigmatic statement in Psalms (90:4) regarding the Divine Day of the biblical Lord:

> *A thousand years, in thy eyes,*
> *[are] as a day past, gone by.*

The statement has traditionally been viewed as merely symbolic of the Lord's eternity. But in view of the numerous traces of Sumerian data in the Book of Psalms (as well as in other parts of the Hebrew Bible), a precise mathematical formula might well have been intended—a formula echoed also in Hindu traditions.

The Hindu traditions were brought to the Indian subcontinent by "Aryan" migrants from the shores of the Caspian Sea, cousins of the Indo-Europeans who were the Hittites of Asia Minor (today's Turkey) and of the Hurrians of the upper Euphrates River, through whom Sumerian knowledge and beliefs were transmitted to the Indo-Europeans. The Aryan migrations are believed to have taken place in the second millennium B.C. and the Vedas were held to be "not of human origin," having been composed by the gods themselves in a previous age. In time the various components of the Vedas and the auxiliary literature that derived from them (the Mantras, Brahmanas, etc.) were augmented by the non-Vedic *Puranas* ("Ancient Writings") and the great epic tales of the Mahabharata and Ramayana. In them, ages deriving from multiples of 3,600 also predominate; thus, according to the *Vishnu Purana,* "the day that Krishna shall depart from Earth will be the first day of the age of Kali; it will continue for 360,000 years of mortals." This is a reference to the concept that the *Kaliyuga,* the present age, is divided to a dawn or "morning twilight" of 100 divine years that equal 36,000 Earth or "mortal" years, the age itself (1,000 divine years equaling 360,000 Earth-years), and a dusk or "evening twilight" of a final 100 divine years (36,000 mortal-years), adding up to 1,200 divine or 432,000 Earth-years.

The depth of such widespread beliefs in a Divine Cycle of 432,000 years, equaling 120 orbits of 3,600 Earth-years each of Nibiru, makes one wonder whether they represent merely arithmetical sleights of hand—or, in some unknown way, a basic natural or astronomical phenomenon recognized in antiquity by the Anunnaki. We have shown in *The 12th Planet,* the first book of *The Earth Chronicles* series, that the Deluge was a global calamity anticipated by the Anunnaki, resulting from the gravitational pull of the nearing Nibiru on the unstable ice sheet over Antarctica. The event brought the last ice age to an abrupt end circa 13,000 years ago, and was thus recorded in Earth's cycles as a major geological and climatic change.

Such changes, the longest being the geological epochs, have been verified through studies of the Earth's surface and oceanic sediments. The last geological epoch, called the Pleistocene, began about 2,500,000 years ago and ended at the time of the Deluge; it was the time span during which hominids evolved, the Anunnaki came to Earth, and Man, *Homo sapiens,* was brought into being. And it was during the Pleistocene that a cycle of approximately 430,000 years was identified in marine sediments. According to a series of studies by teams of geologists led by Madeleine Briskin of the University of Cincinnati, sea level changes and deep-sea climatic records show a "430,000-year quasi-periodic cyclicity." Such a cyclic periodicity conforms with the Astronomical Theory of climatic modulations that takes into account changes due to obliquity (the Earth's tilt), precession (the slight orbital retardation), and eccentricity (the shape of the elliptical orbit). Milutin Milankovitch, who outlined the theory in the 1920s, estimated that the resulting grand periodicity was 413,000 years. His, and the more recent Briskin cycle, almost conform to the Sumerian cycle of 432,000 Earth-years attributed to Nibiru's effects: the convergence of orbits and perturbations and climatic cycles.

The "myth" of Divine Ages thus appears to be based on scientific facts.

The element of Time features in the ancient records, both Sumerian and biblical, not only as a point of beginning—"When." The process of creation is at once linked to the *measurement* of time, measurements that in turn are linked to determinable celestial motions. The destruction of Tiamat and the ensuing creation of the Asteroid Belt and Earth required, according to the Mesopotamian version, two return orbits of the Celestial Lord (the invading Nibiru/Marduk). In the biblical version, it took the Lord two divine "days" to complete the task; hopefully, even Fundamentalists will by now agree that these were not day and night days as we now know them, for the two "days" occurred before Earth had yet come into existence (and besides, let them heed the Psalmist's statement of the Lord's day being equal to a thousand years or so). The Mesopotamian version clearly measures Creation Time or Divine Time by the passages of Nibiru, in an orbit equaling 3,600 Earth-years.

Before that ancient story of Creation shifts to the newly formed Earth and evolution upon it, it is a tale of stars, planets, celestial orbits; and the Time it deals with is *Divine Time*. But once the focus shifts to Earth and ultimately to Man upon it, the scale of Time also shifts—to an *Earthly Time*—to a scale appropriate not only to Man's abode but also to one that Mankind could grasp and measure: Day, Month, Year.

Even as we consider these familiar elements of Earthly Time, it should be borne in mind that all three of them are also expressions of celestial motions—cyclical motions—involving a complex correlation between Earth, Moon, and Sun. We now know that the daily sequence of light and darkness that we call a Day (of twenty-four hours) results from the fact that Earth turns on its axis, so that as it is lit by the Sun's rays on one side, the other side is in darkness. We now know that the Moon is always there, even when unseen, and that it wanes and waxes not because it disappears but because, depending on the Earth-Moon-Sun positions (Fig. 2) we see the Moon fully lighted by the Sun's rays, or fully obscured by the Earth's shadow, or in phases in between. It is this threefold relationship that extends the actual orbital period of the Moon around the Earth from about 27.3 days (the "sidereal month") to the observed cycle of about 29.53 days (the "synodic month") and the

Sun's Rays

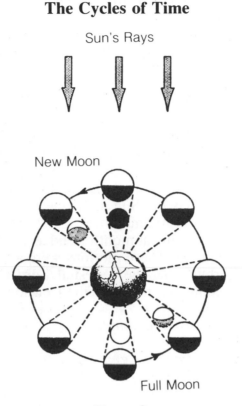

Figure 2

phenomenon of the reappearing or New Moon with all its calendrical and religious implications. And the year or Solar Year, we now of course know, is the period it takes the Earth to complete one orbit around the Sun, our star.

But such basic truths regarding the causes of the Earthly Time cycles of day, month, year are not self-evident and required advanced scientific knowledge to be realized. For the better part of two thousand years it was believed, for example, that the day–night cycle resulted from the circling of Earth by the Sun; for from the time of Ptolemy of Alexandria (second century A.D.) until the "Copernican Revolution" in 1543 A.D., the unquestioned belief was that the Sun, the Moon, and the visible planets were circling the Earth, which was the center of the universe. The suggestion by Nicolaus Copernicus that the Sun was at the center and that the Earth was just another celestial body orbiting it, like any other planet, was so revolutionary scientifically and heretical religiously that he delayed writing his great astronomical work (*De revolutionibus coelestium;* English translation, *On the Revolutions of Celestial Spheres*) and his friends delayed printing it until his very last day, May 24, 1543.

Yet it is evident that in earlier times Sumerian knowledge included familiarity with the triple Earth-Moon-Sun relationship. The *Enuma elish* text, describing the four phases of the Moon, clearly explained them in terms of the position of the Moon vis-a-vis the Sun as it (the Moon) circled the Earth: a full moon at midmonth as it "stood still opposite the Sun," and its waning at month's end as it "stood against the Sun" (see Fig. 2). These motions were attributed to the "destinies" (orbits) that the Celestial Lord (Nibiru) gave Earth and its moon as a result of the Celestial Battle:

The Moon he caused to shine,
to it the night entrusting;
In the night the days to signal
he appointed it, [saying:]
Monthly, without cease, form designs with a crown.
At the month's very start, rising over the Earth,
thou shalt have luminous horns to signify six days,
reaching a crescent on the seventh day.
At mid month stand still opposite the Sun;
it shall overtake thee at the horizon.
Then diminish thy crown and regress in light,
at that time approaching the Sun;
And on the 30th day thou shalt stand against the Sun.
I have appointed thee a destiny; follow its path.

"Thus," the ancient text concludes, did the Celestial Lord "appoint the days and establish the precincts of night and day."

(It is noteworthy that the biblical and Jewish tradition, according to which the twenty-four-hour day begins at sundown the previous evening—"and it was *evening* and it was morning, one day"—is already expressed in the Mesopotamian texts. In the words of *Enuma elish,* the Moon was "appointed *in the night* the days to signal.")

Even in its condensed version of the much more detailed Mesopotamian texts, the Bible (Genesis 1:14) expressed the triple relationship between Earth, Moon, and Sun as it applied to the cycles of day, month, year:

> *And the Lord said:*
> *Let there be luminaries*
> *in the hammered-out Heaven*
> *to distinguish between the day and the night;*
> *And let them be signs*
> *for months and for days and for years.*

The Hebrew term *Mo'edim* used here to denote "months," which signifies the ritual assembly called for on the evening of the New Moon, establishes the Moon's orbital period and phases as an integral component of the Mesopotamian-Hebrew calendar from its very inception. By listing the two luminaries (Sun and Moon) as responsible for the months and the days and the years, the complex lunar-solar nature of that calendar's antiquity is also presented. Over the millennia of Mankind's efforts to measure time by devising a calendar, some (as the Moslems continue to this day) have followed only the Moon's cycles; others (as the ancient Egyptians and the Common Era calendars in use in the Western world) have adopted the solar year, conveniently dividing it into "months." But the calendar devised about fifty-eight hundred years ago in Nippur (Sumer's religious center) and still adhered to by the Jews retained the biblically stated complexity of time-keeping based on the orbital relationship between the Earth and the two luminaries. In doing that, the fact that the Earth orbits the Sun was recognized by the term *Shanah* for "year" which stems from the Sumerian *shatu*, an astronomical term meaning "to course, to orbit," and the full term *Tekufath ha-Shanah*—"the circling or annual orbiting" to denote the passage of a full year.

Scholars have been puzzled by the fact that the *Zo'har* (The Book of Splendor), an Aramaic-Hebrew composition which is a central work in the literature of Jewish mysticism known as *Kabbalah*, unmistakably explained—in the thirteenth century of the Christian era—that the cause of the day's changing into night was the turning of the Earth around its own axis. Some two hundred fifty years *before* Copernicus asserted that the day–night sequence resulted not from the Sun's circling of the Earth but from the Earth's turning on its own axis, the *Zohar* stated that "The entire Earth spins, turning as a sphere. When one part is down the other part is up. When it is light for one part it is dark for the other part; when it is day for that, it is night for the other." The *Zohar's* source was the *third century* Rabbi Hamnuna!

Though little known, the role of Jewish savants in transmitting astronomical knowledge to Christian Europe in the Middle Ages has been convincingly documented by extant books on astronomy, written in Hebrew and containing clear illustrations (as this one from a twelfth century book published in Spain, Fig. 3). Indeed, the writings of Ptolemy of Alexandria, known to the Western world as the *Almagest*, were first preserved by the Arab conquerors of Egypt in the eighth century and became available to Europeans through translations by Jewish scholars; significantly some of these translations contained commentaries casting doubt on the

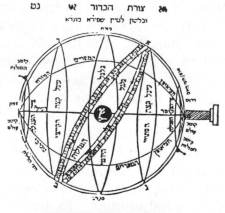

Figure 3

accuracy of the geocentric theories of Ptolemy centuries before Copernicus. Other such translations of Arabic and Greek works on astronomy, as well as independent treatises, were a main channel for the study of astronomy in medieval Europe. In the ninth and tenth centuries Jewish astronomers composed treatises on the movements of the Moon and the planets and calculated the paths of the Sun and the positions of the constellations. In fact, the compilation of astronomical tables, whether for European kings or Moslem caliphs, was a specialty of Jewish court astronomers.

Such advanced knowledge, seemingly ahead of its time, can be explained only by the retention of the earlier sophisticated knowledge that permeates the Bible and its earlier Sumerian sources. Indeed, *Kabbalah* literally means "that which was received," earlier secret knowledge transmitted from generation to generation. The knowledge of Jewish savants in the Middle Ages can be traced directly to academies in Judea and Babylonia that commented upon and retained biblical data. The *Talmud,* recording such data and commentaries from about 300 B.C. to about A.D. 500, is replete with astronomical snippets; they include the statement that Rabbi Samuel "knew the paths of heaven" as if they were the streets of his town, and the reference by Rabbi Joshua ben-Zakai to "a star which appears once in seventy years and confounds the mariners"—familiarity with Halley's Comet whose periodic return every seventy-five years or so was assumed to have been unknown until discovered by Edmund Halley in the eighteenth century. Rabbi Gamliel of Jabneh possessed a tubular optical instrument with which he observed the stars and planets—fifteen centuries before the "official" invention of the telescope.

The need to know the heavenly secrets stemmed from the lunar-solar nature of the Jewish (i.e. Nippurian) calendar, which required a complex adjustment—"intercalation"—between the solar year and the lunar year, the latter falling short of the former by 10 days, 21 hours, 6 minutes and about 45.5 seconds. That shortfall equals $7/19$ of a synodic month, and, therefore, a lunar year can be realigned with the solar year by adding seven lunar months to every nineteen solar years. Astronomy books credit the Athenian astronomer Meton (circa 430 B.C.) with the discovery of this nineteen-year cycle; but the knowledge in fact goes back millennia, to ancient Mesopotamia.

Scholars have been puzzled by the fact that in the Sumerian-Mesopotamian pantheon, Shamash (the "Sun god") was depicted as the son of the "Moon god" Sin, and thus of a lesser hierarchical standing, rather than the expected reverse order. The explanation may lie in the origins of the calendar, wherein the notation of the cycles of the Moon preceded the measurement of the solar cycle. Alexander Marshack, in *The Roots of Civilization*, suggested that markings on bone and stone tools from Neanderthal times were not decorations but primitive lunar calendars.

In the purely lunar calendars, as is still the case in the Moslem calendar, the holidays keep slipping back by about a month every three years. The Nippurian calendar, having been devised to maintain a cycle of holidays connected with the seasons, could not allow such an ongoing slippage: the New Year, for example, had to begin on the first day of spring. This required, from the very beginning of Sumerian civilization, a precise knowledge of the motions of the Earth and the Moon, and their correlation with the Sun, and thus the secrets of intercalation. It also required understanding how the seasons come about.

Nowadays we know that the annual movement of the Sun from north to south and back, causing the seasons, results from the fact that the Earth's axis is tilted relative to the plane of its orbit around the Sun; this "obliquity" is at present about 23.5 degrees. The farthest points reached by the Sun north and south, where it seems to hesitate, then turn back, are called solstices (literally, "Sun standstills"), occurring on June 21 and December 22. The discovery of the solstices has also been attributed to Meton and his colleague, the Athenian astronomer Euctemon. But, in fact, such knowledge goes back to much earlier times. The rich astronomical vocabulary of the Talmud had already applied the term *Neti'yah* (from the verb *Natoh*, "to tilt, incline, turn sideways") to the modern equivalent term "obliquity";

a millennium earlier the Bible recognized the notion of the Earth's axis by attributing the day–night cycle to a "line" drawn through the Earth (Psalms 19:5); and the Book of Job, speaking of the formation of the Earth and its mysteries, attributed to the Celestial Lord the creation of an inclined line, a tilted axis, for the Earth (Job 38:5). Using the term *Natoh,* the Book of Job refers to the Earth's tilted axis and the North Pole when it states (26:7)

> *He tilted north over the void*
> *and hangeth the Earth upon nothing at all.*

Psalms 74:16–17 recognized not only the correlation between the Earth, Moon, and Sun, and the Earth's rotation about its axis as the cause of day, night, and the seasons, but also recognized the outermost points, the "limits" of the Sun's apparent seasonal movements, that we call solstices:

> *Thine is the day*
> *and thine also is the night;*
> *the Moon and Sun thou didst ordain.*
> *All the Earth's limits thou hast set,*
> *summer and winter didst create.*

If a line is drawn between the sunrise and sunset points for each solstice, the result is such that the two lines cross above the viewer's head, forming a giant X that divides the Earth, and the skies above it, into four parts. This division has been recognized in antiquity and is referred to in the Bible when it speaks of the "four corners of the Earth" and the "four corners of the skies." The resulting division of the circle of the Earth and the skies into four parts that look like triangles rounded at their bases created for the ancient peoples the image of "wings." The Bible thus spoke of the "four wings of the Earth" as well as of the "four wings of the skies."

A Babylonian map of the Earth, from the first millennium B.C., illustrated this concept of four "corners of the Earth" by literally depicting four ''wings'' attached to the circular Earth (Fig. 4).

The Sun's apparent movement from north to south and back resulted not only in the two clearly opposite seasons of summer and winter, but also the interim seasons of autumn and spring. The latter were associated with the equinoxes, when the Sun passed over the Earth's equator (once going, once coming back)—times at which daylight and nighttime are equal. In ancient Mesopotamia, the New Year began on the day of the spring equinox—the first day of the First Month (*Nisannu*—Month ''when the sign is given''). Even when, at the time of the Exodus, the Bible (Leviticus chapter 23) decreed that the New Year be celebrated on the day of the autumnal equinox, that designated month (Tishrei) was called ''the seventh month,'' recognizing that Nisannu has been the first month. In either case, the knowledge of the equinoxes, attested to by the New Year days, clearly extends back to Sumerian times.

The fourfold division of the solar year (two solstices, two equinoxes) was combined in antiquity with the lunar motions to create the first known formal calendar, the lunar-solar calendar of Nippur. It was used by the Akkadians, Babylonians, Assyrians, and other nations after them, and remains in use to this very day as the Jewish calendar.

Figure 4

For Mankind, Earthly Time began in 3760 B.C.; we know the exact date because, in the year 1992 of the Common Era, the Jewish calendar counts the year 5752.

Between Earthly Time and Divine Time there is Celestial Time.

From the moment Noah stepped out of the ark, needing reassurance that the watery end of all flesh would not soon recur, Mankind has lived with a lingering notion—or is it a recollection?—of cycles or eons or Ages of Earth's destruction and resurrection, and has looked to the heavens for celestial signs, omens of good or bad to come.

From its Mesopotamian roots the Hebrew language retains the term *Mazal* as meaning "luck, fortune" which could be either good or bad. Little is it realized that the term is a celestial one, meaning zodiac house, and harkens back to the time when astronomy and astrology were one and the same, and priests atop temple-towers followed the movements of the Celestial Gods to see in which house of the zodiac—in which *Manzalu,* in Akkadian—they stood that night.

But it was not Man who had first grouped the myriads of stars into recognizable constellations, defined and named those that spanned the ecliptic, and divided them into twelve to create the twelve houses of the zodiac. It was the Anunnaki who had conceived of that for their own needs; Man adopted that as his link, his means of ascent, to the heavens from the mortality of life on Earth.

For someone arriving from Nibiru with its vast orbital "year" on a fast orbiting planet (Earth, the "seventh planet" as the Anunnaki had called it) whose year is but one part of 3,600 of theirs, time-keeping had to pose a great problem. It is evident from the Sumerian King Lists and other texts dealing with the affairs of the Anunnaki that for a long time—certainly until the Deluge—they retained the *sar,* the 3,600 Earth-years of Nibiru, as the divine unit of time. But what could they do somehow to create a reasonable relationship, other than 1:3600, between that Divine Time and the Earthly Time?

The solution was provided by the phenomenon called precession. Because of its wobble, the Earth's orbit around the Sun is slightly retarded each year; the retardation or precession amounts to 1° in seventy-two years. Devising the division of the ecliptic (the plane of planetary orbits around the Sun) into twelve—to conform to the twelve-member composition of the Solar System—the Anunnaki invented the twelve houses of the zodiac; that allotted to each zodiac house 30°, in consequence of which the retardation per house added up to 2,160 years (72 × 30 = 2,160) and the complete Precessional Cycle or "Great Year" to 25,920 years (2,160 × 12 = 25,920). In *Genesis Revisited* we have suggested that by relating 2,160 to 3,600 the Anunnaki arrived at the Golden Ratio of 6:10 and, more importantly, at the sexagesimal system of mathematics which multiplied 6 by 10 by 6 by 10 and so on and on.

"By a miracle that I have found no one to interpret," the mythologist Joseph Campbell wrote in *The Masks of God: Oriental Mythology* (1962), "the arithmetic that was developed in Sumer as early as c. 3200 B.C., whether by coincidence or by intuitive induction, so matched the celestial order as to amount in itself to a revelation." The "miracle," as we have since shown, was provided by the advanced knowledge of the Anunnaki.

Modern astronomy, as well as modern exact sciences, owes much to the Sumerian "firsts." Among them the division of the skies about us and all other circles into 360 portions ("degrees") is the most basic. Hugo Winckler, who with but a few others combined, at the turn of the century, mastery of "Assyriology" with knowledge of astronomy, realized that the number 72 was fundamental as a link between "Heaven, Calendar and Myth" (*Altorientalische Forschungen*). It was so through the *Hameshtu,* the "fiver" or "times five," he wrote, creating the fundamental number 360 by multiplying the celestial 72 (the precessional shift of 1°) by the human 5 of an Earthling's hand. His insight, understandably for his time, did not lead him to envision the role of the Anunnaki, whose science was needed to know of Earth's retardation to begin with.

Among the thousands of mathematical tablets discovered in Mesopotamia, many that served as ready-made tables of division begin with the astronomical number 12,960,000 and end with 60 as the 216,000th part of 12,960,000. H.V. Hilprecht (*The Babylonian Expedition of the University of Pennsylvania*), who studied thousands of mathematical tablets from the library of the Assyrian king Ashurbanipal in Nineveh, concluded that the number 12,960,000 was literally astronomical, stemming from an enigmatic Great Cycle of 500 Great Years of complete precessional shifts (500 × 25,920 = 12,960,000). He, and others, had no doubt that the phenomenon of precession, presumably first mentioned by the Greek Hipparchus in the second century B.C., was already known and followed in Sumerian times. The number, reduced by ten to 1,296,000, it will be recalled, appears in Hindu tradition as the length of the Age of Knowledge as a threefold multiple of the cycle of 432,000 years. The cycles-within-cycles, interplaying 6 and 12 (the 72 years of a 1° zodiacal shift), 6 and 10 (the ratio of 2,160 and 3,600) and 432,000 to 12,960,000, may thus reflect small and great cosmic and astronomical cycles—secrets yet to be unveiled, of which Sumerian numbers offer just a glimpse.

The selection of the vernal equinox day (or conversely, the autumnal equinox day) as the moment to begin the New Year was not accidental, for because of the Earth's tilt, it is just on these two days that the Sun rises at the points where the celestial equator and the ecliptic circle intersect. Because of precession—the full term is Precession of the Equinoxes—the zodiacal house in which this intersection occurs keeps shifting back, appearing in a preceding 1° in the zodiacal band every seventy-two years. Although this point is still being referred to as the First Point of Aries, in fact we have been in the "Age" (or zodiac) of Pisces since about 60 B.C., and slowly but surely we will soon enter the Age of Aquarius (Fig. 5). It is such a shift—the change from a fading zodiacal age to the start of another zodiacal age—that is the coming of a *New Age*.

As Mankind on Earth awaits the change with anticipation,

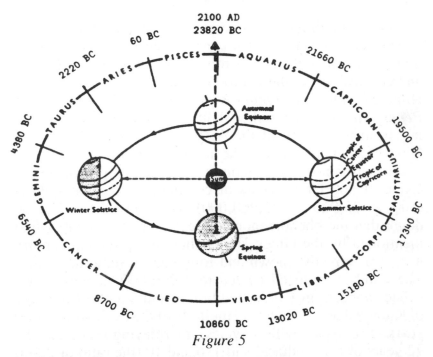

Figure 5

many are those who wonder what the change will bring with it—of what *Mazal* will it be a harbinger? Bliss or upheavals, an end—or a new beginning? The end of the Old Order or the start of a New Order on Earth, perhaps the prophesied return of the Kingdom of Heaven to Earth?

Does Time only flow forward or can it also flow backward, philosophers have wondered. In fact, Time does shift backward, for that is the essence of the phenomenon of precession: the retardation in Earth's orbit around the Sun that causes, once in about 2,160 years, the observance of sunrise on the spring equinox not in the next zodiacal house but in the *preceding* one . . . Celestial Time, as we have designated it, does not progress in the direction of Earthly (and all Planetary) Time, counterclockwise; rather, it moves in the opposite direction, matching the orbital (clockwise) direction of Nibiru.

Celestial Time does flow backward, as far as we on Earth are concerned; and therefore, in zodiacal terms, *the Past is the Future*.

Let us examine the Past.

2

A COMPUTER
MADE OF STONE

The notion or recollection of cyclical ages affecting Earth and Mankind was not confined to the Old World. When Hernando Cortés was welcomed by the Aztec king Moctezuma as a returning god, he was presented with an immense golden disk on which were carved the symbols of the cyclical ages in which the Aztecs and their predecessors in Mexico believed. That precious artifact has been lost forever, having been quickly melted down by the Spaniards; but replicas thereof, in stone, have been found (Fig. 6). The glyphs represented the cycle of "Suns" or ages of which the present is the fifth. The previous four all ended in one or another natural calamity—water, wind, quakes and storms, and wild animals. The first age was the Age of the White Haired Giants; the second, the Golden Age. The third

Figure 6

27

was the Age of the Red Haired People (who, according to the legends, were the first to arrive by ships in the Americas); and the fourth was the Age of the Black Haired People, with whom the supreme Mexican god, Quetzalcoatl, had arrived.

All the way south in pre-Columbian Peru, the Andean peoples also spoke of five "Suns" or ages. The first one was the age of the *Viracochas,* white and bearded gods; the second was the Age of the Giants, followed by the Age of Primitive Man. The fourth was the Age of Heroes; and then came the fifth or contemporary age, the Age of Kings, of which the Inca kings were last in line. The durations of these ages were measured in thousands rather than in tens or hundreds of thousands of years. Mayan monuments and tombs were decorated with "sky bands" whose glyphs have been found to represent the zodiacal division of the heavens; artifacts found in Mayan ruins and in the Inca capital Cuzco have been identified as zodiacal calendars. The city of Cuzco itself, it appears, was (in the words of S. Hagar in a paper delivered at the 14th Congress of Americanists) "a testimonial in stone" to the South American familiarity with the twelve-house zodiac. The unavoidable conclusion is that knowledge of the zodiacal division of the ecliptic was somehow known in the New World millennia ago, and that the Ages were measured in the 2,160-year units of Celestial Time.

The idea that calendars could be made of stone might seem strange to us, but was evidently quite logical in antiquity. One such calendar, posing many puzzles, is called *Stonehenge.* It consists nowadays of gigantic stone blocks that stand silently on a windswept plain in England, north of the city of Salisbury and about eighty miles southwest of London. The remains pose an enigma that has titillated the curiosity and imagination of generations, challenging historians, archaeologists, and astronomers. The mystery these megaliths bespeak is lost in the mists of earlier times; and Time, we believe, is the key to its secrets.

Stonehenge has been called "the most important prehistoric monument in the whole of Britain," and that alone justifies the attention it has been given over the centuries and especially in recent times. It has been described—at least by its British relators—as unique, for "there is nothing else like it anywhere in the world" (R.J.C. Atkinson, *Stonehenge and Neighbouring Monuments*); and that may explain why an eighteenth-century manuscript listed more than six hundred works on Stonehenge in its catalogue of ancient monuments in Western Europe. Stonehenge is indeed the largest and most elaborate of more than nine hundred ancient stone, wood, and earthen circles in the British Isles, as well as the largest and most complicated one in Europe.

Yet, in our view, it is not only what makes Stonehenge unique that is its most important aspect. It is also what reveals its similarity to certain monuments elsewhere, and its *purpose* at the specific *time* of its construction, that make it part of the tale we have called *The Earth Chronicles*. It is within such a wider framework, we believe, that one can offer a plausible solution to its enigma.

Even those who have not visited Stonehenge must have seen, in print or on the screen, the most striking features of this ancient complex: the pairs of huge upright stone blocks, each about thirteen feet high, connected at the top by an equally massive lintel stone to form freestanding *Trilithons;* and these, erected in a semicircle, surrounded in turn by a massive circle of similar giant stones connected at the top by lintels that were carefully carved to form a continuous ring around the paired uprights. Though some of the stone blocks in what are called the sarsen trilithons and the *Sarsen Circle* (after the type of stone, a kind of sandstone, to which these boulders belong) are missing and some have toppled, it is they that create the view that the word "Stonehenge" conjures (Fig. 7).

Inside this massive stone ring other, smaller stones called bluestones were placed so as to form the *Bluestone Circle* outside the Trilithons and a bluestone semicircle (some refer to it as the *Bluestone Horseshoe*) inside the Trilithon half-circle. As is the case regarding the sarsen stones, not all of the bluestones that together formed these circles and half-circles (or "horseshoes") are still in place. Some are miss-

Figure 7

ing altogether; some lie about as fallen giants. Adding to the site's haunting aura are other gigantic stones that lie about and whose nicknames (of uncertain origin) compound the mystery; they include the *Altar Stone,* a sixteen-foot-long dressed block of blue-gray sandstone that remains half-buried under an upright and the lintel of one of the Triliths. In spite of considerable restoration work, much of the structure's past glory is either gone or fallen. Still, archaeologists have been able to reconstruct from all the available evidence how this remarkable stone monument looked in its prime.

They have concluded that the outer ring, of uprights connected by curved lintels, consisted of thirty upright stones of which seventeen remain. Within this Sarsen Circle there stood the Bluestone Circle of smaller stones (of which twenty-nine are still extant). Within this second ring stood five pairs of Trilithons, making up the *Sarsen Horseshoe* of ten massive sarsen blocks; they are usually numbered 51 through 60 on charts (lintel stones are numbered separately in a series that adds 100 to their related uprights; thus the lintel connecting the uprights 51–52 is number 152).

The innermost semicircle consisted of nineteen bluestones (some numbered 61–72), forming the so-called Bluestone Horseshoe; and within this innermost compound, precisely on the axis of the whole Stonehenge complex, stood the so-called Altar Stone, giving these circles within circles of stone the layout envisioned in Fig. 8a.

As if to emphasize the importance of the circular shape already evident, the rings of stones are in turn centered within a large framing circle. It is a deep and wide ditch whose excavated soil was used to raise its banks; it forms a perfect encompassing ring around the whole Stonehenge complex, a ring with a diameter in excess of three hundred feet. Approximately half the circuit of the ditch was excavated earlier this century and then partly refilled; the other portions of the ditch and its raised banks bear the marks of being weathered down by nature and man over the millennia.

These circles within circles have been repeated in yet other ways. A few feet away from the inner bank of the ditch there exists a circle made up of fifty-six pits, deep and perfectly dug into the ground, called the *Aubrey Holes* after their seventeenth-century discoverer, John Aubrey. Archaeologists have excavated these holes for whatever clues the accumulation of debris might disclose about the site and its builders, and have thereafter plugged up the holes with white cement discs; the result is that the perfect circle that these holes form stands out—especially from the air. In addition, cruder and more irregular holes were dug at some unknown time in two circles around the sarsen and bluestone circles, now known as the Y and Z holes.

Two stones, unlike all the others, have been found positioned on opposite sides of the ditch's inner embankment; and somewhat farther down the line of the Aubrey Holes (but evidently not part of them), two circular mounds, equidistant from the two stones, have been found with holes in them. Researchers are convinced that the holes also held stones akin to the first two, and that the four—called *Station Stones* (now numbered 91–94)—served a distinct purpose, especially since, when connected by lines, the four stones outline a perfect rectangle with probable astronomical connotations. Yet another massive stone block, nicknamed the *Slaughter Stone,* lies fallen where the embanked ditch has a wide gap that clearly served as the opening into (or from) the concentric rings of stones, holes, and earthworks. It probably lies not exactly where it once stood, and was probably not alone, as holes in the ground suggest.

The opening in the ditch is oriented exactly to the northeast. It leads to (or allows arrival from) a causeway, called the *Avenue*. Two parallel embanked ditches outline this avenue, leaving a clear passage over thirty feet wide. It runs straight for more than a third of a mile where it branches northward toward a vast elongated earthwork known as the *Cursus*, whose orientation is at an angle to that of the Avenue; the other branch of the Avenue curves toward the River Avon.

The concentric circles of Stonehenge with the Avenue leading to the northeast (Fig. 8b) provide a major clue re-

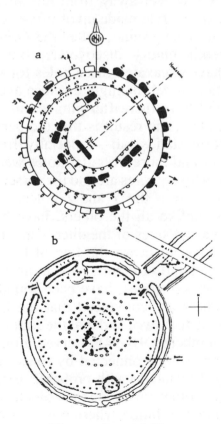

Figures 8a and 8b

garding the purpose for which Stonehenge was constructed. That the direction of the Avenue—its precise northeastern orientation—was not accidental becomes clear when it is realized that a line drawn through the center of the Avenue passes through the center of the circles of stones and holes to form the structure's axis (see Fig. 8a). That the axis was deliberately oriented is suggested by a series of holes indicating that the marker stones had once been placed along this axis. One of them, called the *Heel Stone,* still stands as a mute witness to the builders' intentions and the site's purpose; it was undoubtedly astronomical.

The idea that Stonehenge was a carefully planned astronomical observatory rather than a heathen cult or occult site (a notion expressed, for example, by calling a fallen stone "Slaughter Stone," implying human sacrifices), was not easily accepted. In fact, the difficulty grew rather than diminished the more the site was investigated and its date of construction kept shifting backward.

A twelfth-century account (*Historia regum Britanniae* by Geoffrey of Monmouth) related that the "Giants' Ring" was "a stone cluster which no man of the period could ever erect and was first built in Ireland from stones brought by the giants from Africa." It was then on the advice of the sorcerer Merlin (whom Arthurian legends also connected with the Holy Grail) that the King of Vortigen moved the stones and "re-erected them in a circle round a sepulchre, in exactly the same way as they had been arranged on Mount Killaraus" in Ireland. (That this medieval legend had a factual core was given confirmation by the modern discovery that the bluestones originated from the Prescelly Mountains in southwestern Wales and were somehow transported by land and water over a distance of two hundred fifty miles— first to a site some twelve miles northwest of Stonehenge, where they might have been erected in an earlier circle, and then on to Stonehenge proper).

In the seventeenth and eighteenth centuries, the stone temple was attributed to the Romans, the Greeks, the Phoenicians, or the Druids. The common aspect of these various

notions is that they all shifted the time attributed to Stonehenge from the Middle Ages back to the beginning of the
Christian era and earlier, thus substantially increasing the
site's antiquity. Of these various theories, the one concerning the Druids gained the most favor at the time, not least
of all because of the research and writings of William Stukeley, especially his 1740 work *Stonehenge, A Temple Re-
stor'd To The British Druids*. The Druids were the learned
class or sect of teacher-priests among the ancient Celts.
According to Julius Caesar, who is the prime source of
information regarding the Druids, they assembled once a
year at a sacred place for secret rites; they offered human
sacrifices; and among the subjects they taught the Celt no-
blemen were "the powers of the gods," the sciences of
nature, and astronomy. While nothing that has been uncovered by archaeologists at the site reveals any connection
with pre–Christian era Druids, the Celts had arrived in the
area by that time and there is no proof the other way either,
namely that the Druids did not gather at this "Sun Temple"
even if they had nothing to do with its much earlier builders.

Although Roman legions encamped near the site, no evidence was found to connect Stonehenge with the Romans.
A Greek and Phoenician connection, however, shows more
promise. The Greek historian Diodorus Siculus (first century
B.C.)—a contemporary of Julius Caesar—who had traveled
to Egypt, wrote a multivolume history of the ancient world.
In the first volumes he dealt with the prehistory of the
Egyptians, Assyrians, Ethiopians, and Greeks, the so-called
"mythic times." Drawing on the writings of earlier historians, he quotes from a (by now lost) book by Hecataeus
of Abdera in which the latter had stated, circa 300 B.C.,
that on an island inhabited by the Hyperboreans "there is
a magnificent sacred precinct of Apollo and a notable temple
which is spherical in shape." The name in Greek signified
a people from the distant north, whence the north wind
("Boreas") comes. They were worshipers of the Greek
(later Roman) god Apollo, and the legends regarding the
Hyperboreans were thus mingled with the myths concerning

Apollo and his twin sister, the goddess Artemis. As the ancients told it, the twins were the children of the great god Zeus and their mother Leto, a Titaness. Impregnated by Zeus, Leto wandered over the face of the Earth seeking a place to give birth to her children in peace, away from the wrath of Hera, the official wife of Zeus; Apollo was thus associated with the distant north. The Greeks and the Romans considered him a god of divination and prophecy; he circled the zodiac in his chariot.

Though not attributing any scientific value to such a legendary or mythological connection with Greece, archaeologists have nevertheless seemed to find such a connection through archaeological discoveries in the area of Stonehenge, which is replete with prehistoric earthworks, structures, and graves. These man-made ancient remains include the great Avebury Circle, which schematically drawn resembles the works of a modern watch (Fig. 9a, as sketched by William Stukeley) or even the meshing wheels of the ancient Mayan calendar (Fig. 9b). They also include the miles-long trench called the Cursus; a kind of wooden-pegged rather than stone-made circle called Woodhenge; and the outstanding Silbury Hill—an artificial conical hill

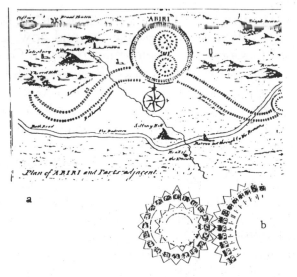

Figures 9a and 9b

which is precisely circular and 520 feet in diameter, the largest of its kind in Europe (some see significance in the fact that it is situated an exact six purported "megalithic miles" from Stonehenge).

The most important finds, archaeologically speaking, have been made in this area (as often elsewhere) in tombs, which are scattered all over the Stonehenge area. In them archaeologists have found bronze daggers, axes and maces, golden ornaments, decorated pottery, and polished stones. Many of those finds reinforced archaeological opinion that the manner in which stones at Stonehenge were smoothly dressed and carefully shaped indicated "influences" from Minoan Crete (the Mediterranean island) and Mycenaean (mainland) Greece. It was also noted that some of the peg-into-socket joints used at Stonehenge to hold together stone blocks were similar to the joints used in the stone gateways of Mycenae. All this, many archaeologists held, pointed to a connection with ancient Greece.

A leading representative of this school has been Jacquetta Hawkes, who in her book, *Dawn of the Gods,* about the Minoan and Mycenaean origins of Greek civilization, could not help devoting a good portion of the chapter on "Graves and Kingdoms" to Stonehenge.

Mycenae is situated in the southwestern part of mainland Greece that is called the Peloponnesus (and now separated from the rest of Greece by a man-made Corinth Canal) and acted as a bridge between the earlier Minoan civilization on the island of Crete and the later classical Greek one. It flowered in the sixteenth century B.C. and the treasures uncovered in the tombs of its kings revealed foreign contacts that undoubtedly included Britain. "At just this time when Mycenaean kings were rising to a new wealth and power," Jacquetta Hawkes wrote, "a rather similar advance, although on a smaller scale, was taking place in southern England. There too a warrior aristocracy was ruling over peasants and herders and beginning to trade and to prosper— and to be buried with appropriate extravagance. Among the possessions so buried were a few objects that prove these chieftains to have had contacts with the Mycenaean world."

Such things, she added, were not of great moment and could just be the fruits of trade or imitation, were it not for "the unique event—the building of the great sarsen-stone circle and trilithons of Stonehenge."

Not all archaeological finds, however, showed such early Greek "influences." The finds in tombs around Stonehenge included, for example, decorated beads and amber disks bound with gold in a method developed in Egypt and not at all in Greece. Such finds raised the possibility that all those artifacts were somehow imported to southeast England, neither by Greeks nor Egyptians but perhaps by trading people from the eastern Mediterranean. The obvious candidates were the Phoenicians, the renowned sailors-cum-traders of antiquity.

It is a recorded fact that the Phoenicians, sailing from their Mediterranean ports, reached Cornwall in the southwest corner of England, quite close to Stonehenge, in the search for tin, with which hardened bronze was made from soft copper. But were any of these peoples, whose trade links flourished in the millennium between 1500 B.C. and 500 B.C., responsible for the planning and construction of Stonehenge? Did they even visit it? A partial answer would depend, of course, on when Stonehenge itself was conceived and built, or who else was there to build it.

In the absence of written records or carved images of the Mediterranean gods (artifacts found elsewhere among Minoan, Mycenaean, and Phoenician ruins) no one can answer the question with any certainty. But the question itself became moot when various remains of organic origin, such as carved antlers, were dug up by archaeologists at Stonehenge. Subjected to radiocarbon dating, remains found in the Ditch produced a date of between 2900 to 2600 B.C.—at least a thousand years and probably much more before the sailors from the Mediterranean may have arrived. A charcoal piece found in one of the Aubrey Holes provided a carbon date of 2200 B.C.; an antler pick found near one of the trilithons gave a reading of between 2280 and 2060 B.C.; radiocarbon datings of finds in the Avenue gave dates between 2245 and 2085 B.C.

Who was there at such an early time to plan and execute the marvelous stone complex? Scholars hold that until about 3000 B.C. the area was sparsely populated by small groups of early farmers and herders who used stone for their tools. Some time after 2500 B.C. new groups arrived from the European continent; they brought with them knowledge of metals (copper and gold), used clay utensils, and buried their dead in round mounds; they have been nicknamed the Beaker People, after the shape of their drinking vessels. At about 2000 B.C. bronze made its appearance in the area and a wealthier and more numerous population, known as the Wessex People, engaged in cattle ranching, metal crafts, and trade with western and central Europe and the Mediterranean. By 1500 B.C. this era of prosperity suffered an abrupt decline that lasted the better part of a millennium; and Stonehenge must have shared in this decline.

Were the Neolithic farmers and herders, the Beaker People, or even the Early Bronze Age Wessex People, capable of creating Stonehenge? Or did they just provide the labor and the manpower to construct a complex mechanism in stone devised by advanced scientific knowledge of others?

Even an outspoken proponent of the Mycenaean connection, Jacquetta Hawkes, had to admit that Stonehenge, "this sanctuary, constructed from colossal yet carefully-shaped blocks that make the cyclopean masonry of Mycenae look like children's bricks, has nothing to compare with it in all prehistoric Europe." To allow for the Mycenaean connection and to link it with the early Englanders, she proceeded to offer the theory that "some of the local lords controlling the pastures of Salisbury plain, and perhaps, like Odysseus, owning twelve herds of cattle, may have had the wealth and authority needed to turn what had been a modest sanctuary of Stone Age origin into a noble and unparalleled work of megalithic architecture. It has always seemed that some individual must have initiated it—through swollen ambition or religious obsession—but because the whole design and method of building is so far advanced on anything known in the island before, it has seemed likely that ideas drawn from a more civilized tradition might also have been involved.''

But what was that ''more civilized tradition'' that gave rise to this structure that was beyond compare to anything in prehistoric Europe? The answer must depend on an accurate dating of Stonehenge; and if, as scientific data suggests, it is a thousand to two thousand years older than the Mycenaeans and the Phoenicians, then an earlier source of the ''civilized tradition'' must be sought. If Stonehenge belongs to the third millennium B.C., then the only candidates are those of Sumer and Egypt. When Stonehenge was first conceived, the Sumerian civilization, with its cities, high-rise temples-cum-observatories, writing, and scientific knowledge, was already a thousand years old, and kingship had already flourished in Egypt for many centuries.

For a better answer, we have to put together the knowledge accumulated by now regarding the several phases by which Stonehenge, according to the latest research, came to be.

Stonehenge began with hardly any stones. It began, all are agreed, with the Ditch and its embankment, a great earthen circle with a circumference of 1,050 feet at its bottom; it is about twelve feet wide and up to six feet deep, and thus required digging up a considerable quantity of soil (chalky earth) and arranging it to form the two raised banks. Within this outer ring the circle of 56 Aubrey Holes was made.

The northeastern section of the earthen ring was left undug, to provide an entranceway into the midst of the circle. There, two ''gateway stones,'' now missing, flanked this entry to the enclosure; they also served as focusing aides for the Heel Stone, which was erected on the resultant axis. This massive natural boulder stands sixteen feet above the ground and cuts four feet into the earth; it has been set up inclined at an angle of 24°. A series of holes at the entrance gap may have been intended to hold movable wooden markers, and are thus called Post Holes. Finally, the four rounded Station Stones were positioned to form a perfect rectangle;

and this completed *Stonehenge I*—the earthen ring, the Aubrey Holes, an entranceway axis, seven stones, and some wooden pegs.

Organic remains and stone tools associated with this phase suggest to scholars that Stonehenge I was constructed sometime between 2900 and 2600 B.C.; the date selected by the British authorities is 2800 B.C.

Whoever constructed Stonehenge I, and for whatever purpose, found it satisfactory for several centuries. Throughout the occupation of the area by the Beaker People, no need to change or improve the arrangement of earthwork and stones was indicated. Then, at about 2100 B.C., just before the arrival of the Wessex People (or perhaps coinciding with it) a spate of extraordinary activity burst upon the scene. The main event was the introduction of the bluestones into the makeup of Stonehenge, making *Stonehenge II* a stone "henge" for the first time.

It was no mean feat to haul the bluestones, weighing up to four tons each, across land and over sea and river for a total distance of some two hundred fifty miles. To this day it is not known why these particular dolerite stones were chosen and why such a great effort was made to bring them to the site, directly or with a short interval at a temporary way station. Whatever the precise route was, it is believed that in the end they were brought to the site's proximity up the River Avon, which explains why the Avenue was extended by some two miles at this phase to connect Stonehenge with the river.

At least eighty (some estimate eighty-two) bluestones were brought over. It is believed that seventy-six of them were intended for the holes that made up the two concentric Q and R holes, thirty-eight per circle; the circles appear to have had openings on their west-facing sides.

At the same time a separate larger stone, the so-called Altar Stone, was erected within the circles exactly on the Stonehenge axis, facing the Heel Stone to the northeast. But as the researchers checked the alignment and the position of the outer stones, they discovered to their surprise that the Heel Stone was shifted in this Phase II somewhat eastward (to the right, as one looks from the enclosure's center); simultaneously, two other stones were erected in a row in front of the Heel Stone, so as to emphasize the new line of sight. To accommodate these changes, the entrance to the enclosure was widened on its right (eastern side) by filling up part of the Ditch, and the Avenue too was widened there.

Unexpectedly the researchers realized that the main innovation of Stonehenge II was not the introduction of the bluestones, but *the introduction of a new axis,* an axis somewhat more to the east than the previous one.

Unlike the seven or so centuries of dormancy for Stone henge I, *Stonehenge III* followed Phase II within decades. Whoever was in charge decided to give the complex a monumental scope and permanence. It was then that the huge sarsen stones, weighing forty to fifty tons each, were hauled to Stonehenge from Marlboro Downs, some twenty miles away. It is generally assumed that seventy-seven stones were brought.

As laborious as the transporting of these boulders with an aggregate weight of thousands of tons was, even more daunting must have been the task of setting them up. The stones were carefully dressed to the desired shapes. The lintels were given a precise curvature, given (somehow) protruding pegs exactly where they had to fit into carved-out sockets where stone joined stone; and then all those prepared stones had to be erected in a precise circle or in pairs, and the holding lintels hauled up to be placed on top. How the task, made more difficult by the site's slope, was achieved, no one really knows.

At this time the realigned axis was also given permanence by the erection of two new massive Gateway Stones, replacing the earlier ones. It is believed that the fallen Slaughter Stone could have been one of the two new Gateway Stones.

In order to make room for the Sarsen Circle and the Trilithon Horseshoe or oval, the two circles of bluestones from Phase II had to be completely dismantled. Nineteen of them were used to form the inner Bluestone Horseshoe (now recognized as an open-ended oval) and fifty-nine, it is believed, were intended to be placed in two new circles of holes (Y and Z), surrounding the Sarsen Circle. The Y circle was meant to hold thirty stones and the Z circle twenty-nine. Some of the other stones of the original eighty-two may have been intended to serve as lintels or (as John E. Wood, *Sun, Moon and Standing Stones,* believes) to complete the oval. The Y and Z circles, however, were never erected; instead the bluestones were arranged in one large circle, the Bluestone Circle, with an undetermined number of stones (some believe sixty). Also uncertain is the time when this circle was erected—right away, or a century or two later. Some also think that additional work, mainly on the Avenue, was done about 1100 B.C.

But for all intents and purposes, the Stonehenge we see was planned in 2100 B.C., executed during the following century, and given its final touches circa 1900 B.C. Modern scientific research methods have thus corroborated the findings—astounding at the time, 1880—of the renowned Egyptologist Sir Flinders Petrie, that Stonehenge dated to circa 2000 B.C. (It was Petrie who devised the stones' numbering system still in use).

In the usual course of scientific studies of ancient sites, archaeologists are the first to be on the scene, and others—anthropologists, metallurgists, historians, linguists, and other experts—follow. In the case of Stonehenge, astronomers led the way. This was not only because the ruins were visible above the surface and required no excavation to reveal them, but also because from the very beginning it seemed almost self-evident that the axis line from the center toward the Heel Stone through the Avenue pointed "to the northeast, whereabouts the Sun rises when the days are longest" (to use the words of William Stukeley, 1740)—toward the point in the sky where the Sun rises at the summer solstice (about June 21). Stonehenge was an instrument to measure the passage of time!

After two and a half centuries of scientific progress, this conclusion is still valid. All are agreed that Stonehenge was not a place of residence; nor was it a burial place. Neither palace nor tomb, it was in essence a temple-cum-observatory, as the ziggurats (step-pyramids) of Mesopotamia and ancient America were. And being oriented toward the Sun when it rises in midsummer, it could be called a Temple of the Sun.

With this basic fact undisputable, it is no wonder that astronomers continue to lead the research concerning Stonehenge. Prominent among them, at the very beginning of this century, was Sir Norman Lockyer, who conducted a comprehensive survey of Stonehenge in 1901 and confirmed the summer solstice orientation in his master work *Stonehenge and Other British Stone Monuments*. Since this orientation is satisfied by the axis alone, subsequent researchers began in time to wonder whether the additional complexity of Stonehenge—the diverse circles, ovals, rectangle, markers—might signify that other celestial phenomena besides sunrise at summer solstice and other time cycles have been observed at Stonehenge.

There have been suggestions to that effect in earlier treatises on Stonehenge. But it was only in 1963, when Cecil A. Newham discovered alignments that suggested that equinoxes too could have been observed and even predicted at Stonehenge, that these possibilities were given modern scientific credence.

His most sensational suggestion, however (first in articles and then in his 1964 book *The Enigma of Stonehenge*), was that Stonehenge must have also been a *lunar* observatory. He based this conclusion on examination of the four Station Stones and the rectangle that they form (Fig. 10); he also showed that whoever had intended to give Stonehenge this capability knew where to erect it, for the rectangle and its alignments had to be sited exactly where Stonehenge is.

All this was at first received with extreme doubt and disdain, because lunar observations are considerably more complex than solar ones. The Moon's motions (around the Earth and together with the Earth around the Sun) are not repeated on an annual basis, because, among other reasons, the Moon orbits the Earth at a slight inclination to the Earth's orbit around the Sun. The complete cycle, which is repeated

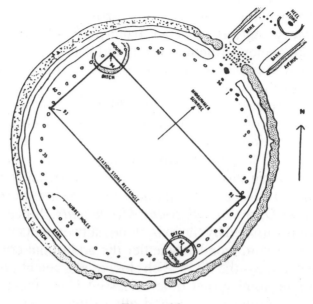

Figure 10

only once in about nineteen years, includes eight points of "Moon Standstill" as the astronomers call them, four major and four minor. The suggestion that Stonehenge I—which already possessed the alignments highlighted by Newham—was built to enable the determination, or even prediction, of these eight points seemed preposterous in view of the fact that Britain's inhabitants at the time were just emerging from the Stone Age. This is clearly a valid argument; and those who have nevertheless found more evidence for the astronomical marvels at Stonehenge are yet to provide an answer to the paradox of a complex lunar observatory amidst Stone Age people!

Prominent among the astronomers whose investigations confirmed the incredible capabilities of Stonehenge was Gerald S. Hawkins of Boston University. Writing in prestigious scientific journals in 1963, 1964, and 1965, he announced his far-reaching conclusions by entitling his studies "Stonehenge Decoded," "Stonehenge: A Neolithic Computer," and "Sun, Moon, Men and Stones," followed by his books *Stonehenge Decoded* and *Beyond Stonehenge*.

With the aid of the university's computers he analyzed hundreds of sight lines at Stonehenge and related them to the positions of the Sun, Moon, and major stars as they were in ancient times, and decided that the resulting orientations could not have been just accidental.

He attached great significance to the four Station Stones and the perfect rectangle they form and showed how the lines connecting opposite stones (91 with 94 and 92 with 93) were oriented to the points of major standstills and those connecting the stones diagonally to the points of minor standstills of the Moon at moonrise and moonset. Together with the four points of the Sun's movements, Stonehenge, according to Hawkins, enables observation and prediction of all the twelve points marking the Sun's and the Moon's movements. Above all he was fascinated by the number 19 expressed by stones and holes in the various circles: the two circles of 38 bluestones of Stonehenge II "can be regarded as two semi-circles of 19" (*Stonehenge Decoded*) and the oval "horseshoe" of Stonehenge III had the exact 19. This was an unmistakable lunar relationship, for 19 was the Moon's cycle which governs intercalation.

Professor Hawkins went even farther: he concluded that the numbers expressed by stones and holes in the various circles bespoke an ability to predict eclipses. Because the Moon's orbit is not exactly in the same plane as the Earth's orbit around the Sun (the former is inclined to the latter by just over 5°), the Moon's orbit crosses the path of the Earth around the Sun at two points each year. The two points of intersection ("nodes") are commonly marked on astronomical charts N and N′; this is when eclipses occur. But because of the irregularities in the shape and lag of the Earth's orbit around the Sun, these nodal intersections do not recur precisely at the same celestial positions year after year; rather, they recur in a cycle of 18.61 years. Hawkins postulated that the operating principle for this cycle was, therefore, "cycle end/cycle start" in the nineteenth year, and Hawkins reasoned that the purpose of the 56 Aubrey Holes was to attain an adjustment by moving three markers at a time within the Aubrey circle, since $18\frac{2}{3} \times 3 = 56$. This, he held, made possible the foretelling of eclipses of the Moon as well as of the Sun, and his conclusion was that such a prediction of eclipses was the main purpose of the construction and design of Stonehenge. Stonehenge, he announced, was nothing short of *a brilliant astronomical computer made of stone*.

The proposition that Stonehenge was not only a "Sun temple" but also a lunar observatory was met at first with fierce resistance. Prominent among the dissenters, who considered many of the Moon alignments to be coincidental, was Richard J.C. Atkinson of the University College in Cardiff, who had led some of the most extensive archaeological excavations at the site. The archaeological evidence for the great antiquity of Stonehenge was the very reason for his disdain for the observatory/lunar-alignments/Neolithic computer theories, for he asserted that Neolithic Man in Britain was simply incapable of such achievements. His disdain and even ridicule, expressed in such titles for his articles in *Antiquity* as "Moonshine on Stonehenge" and in his book *Stonehenge*, turned to grudging support as a result of studies conducted at Stonehenge by Alexander Thom (*Megalithic Lunar Observations*). Thom, an engineering professor at Oxford University, conducted the most accurate measurements at Stonehenge, and pointed out that the "horseshoe" arrangement of the sarsen stones in fact represented an oval (Fig. 11), an elliptical shape that represents

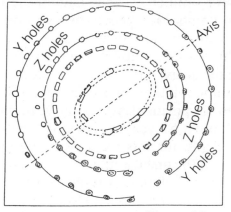

⌒ Upright stones
⊕ Excavated holes
○ Presumed holes

Figure 11

planetary orbits more accurately than a circle. He agreed with Newham that Stonehenge I was primarily a lunar, and not just a solar, observatory, and confirmed that Stonehenge was erected where it is because it is only there that the eight lunar observations could be made precisely along the lines formed by the rectangle connecting the four Station Stones.

The fierce debate, conducted on the pages of leading scientific magazines and in confrontational conferences, was summed up by C.A. Newham (*Supplement to the Enigma of Stonehenge and its Astronomical and Geometric Significance*) in the following words: "With the exception of the five Trilithons, practically all the remaining features appear to have lunar connections." He agreed that the "56 Aubrey Holes rotate to the eight main alignments of the Moon setting and rising." Thereafter, even Atkinson admitted that he "has become sufficiently persuaded that conventional archaeological thinking is in need of drastic revision" in regard to the purpose and functions of Stonehenge.

These conclusions were to no small measure the result of the research by a notable participant who had joined the growing list of involved scientists in the late 1960s and the decade of the 1970s. He was Sir Fred Hoyle, astronomer and mathematician. He held that the alignments listed by Hawkins to various stars and constellations were rather random than deliberate, but fully agreed with the lunar aspects of Stonehenge I—and especially the role of the fifty-six Aubrey Holes and the rectangular arrangement of the Station Stones ("Stonehenge—An Eclipse Predictor" in *Nature* and *On Stonehenge*).

But concurring that the Aubrey Circle could act as a "calculator" for predicting eclipses (in his opinion it was done by moving four markers around), Hoyle stirred up another issue. Whoever had designed this calculator—Hawkins called it a "computer"—must have known *in advance* the *precise* length of the solar year, the Moon's orbital period, and the cycle of 18.61 years; and Neolithic Man in Britain simply did not possess such knowledge.

Struggling to explain how the advanced knowledge of astronomy and mathematics had appeared in Neolithic Britain, Hawkins resorted to ancient records of the Mediterranean peoples. In addition to the Diodorus/Hecataeus reference he also mentioned Plutarch's quote (in *Isis and Osiris*) of Eudoxus of Cnidus, the fourth century B.C. astronomer-mathematician from Asia Minor, who had associated the "demon god of eclipses" with the number fifty-six.

In the absence of answers from Man, a glance at the superhuman?

Hoyle, on his part, arrived at the conviction that Stonehenge was not a mere observatory, a place to see what goes on in the skies. He called it a *Predictor,* an instrument for foretelling celestial events and a facility for noting them on the predetermined dates. Agreeing that "such an intellectual achievement was beyond the capacity of the local Neolithic farmers and herdsmen," he felt that the Station Stones rectangle and all it implied indicate "that the builders of Stonehenge I might have come to the British Isles from the outside, purposely looking for this rectangular alignment" (which is possible just where Stonehenge is located, in the northern hemisphere), "just as the modern astronomer often searches far from home for places to build his telescopes.

"A veritable Newton or Einstein must have been at work at Stonehenge," Hoyle mused; but even so, where was the university where he had learned mathematics and astronomy, where were the writings without which accumulated knowledge could not be passed on and taught, and how could a sole genius plan, execute, and supervise such a celestial predictor when, for Phase II alone, a whole century was needed? "There have only been about 200 generations of history; there were upward of 10,000 generations of prehistory," Hoyle observed. Was it all part of the "eclipse of the gods," he wondered—the transition from a time when people worshiped an actual Sun god and a Moon god, "to become the invisible God of Isaiah?"

Without explicitly divulging his thoughts, Hoyle gave an answer by quoting the full section in Diodorus from Hecataeus regarding the Hyperboreans; it states toward its end that after the Greeks and Hyperboreans exchanged visits "in the most ancient times,"

> They also say that the Moon, as viewed from this island, appears to be but a little distance from the Earth, and to have upon it prominences, like those of the Earth, which are visible to the eye.
> The account is also given that the god visits the island every nineteen years, the period in which the return of the stars to the same place in the heavens is accomplished; and for this reason the nineteen-year period is called by the Greeks the "year of Meton."

The familiarity in such distant times not only with the nineteen-year cycle of the Moon but also with "prominences, like those of the Earth"—surface features such as mountains and plains—is unquestionably amazing.

The attribution by Greek historians of the circular structure in Hyperborea to the lunar cycle first described in Greece by the Athenian Meton tosses the problem of Who Built Stonehenge to the ancient Near East; so do the soul-searching conclusions and musings of the above mentioned astronomers.

But more than two centuries earlier, William Stukeley had already pointed for answers in the same direction, toward the ancient Near East. To his sketch of Stonehenge, as he understood it to have been, he appended the design he had seen on an eastern Mediterranean ancient coin (Fig. 12a) which depicts a temple on an elevated platform. This depiction, more explicit, also appears on another ancient coin from the city of Byblos in the same area, one that we have reproduced in the very first volume of *The Earth Chronicles*. It shows that the ancient temple had an enclosure in which there stood *a rocket upon a launch pad* (Fig. 12b). We have identified the place as The Landing Place of Sumerian lore, the place where the Sumerian king Gilgamesh witnessed a rocket ship rise. The place still exists; it is now the vast platform in the mountains of Lebanon, at Baalbek, upon which there still stand the ruins of the greatest

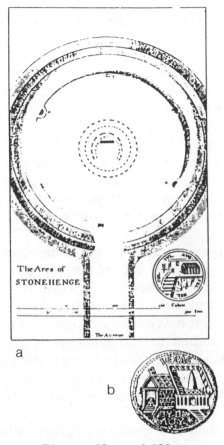

Figures 12a and 12b

Roman temple ever built. Supporting the massive platform are three colossal stone blocks that have been known since antiquity as the *Trilithon*.

The answers to the Stonehenge enigma should thus be sought in places far away from it, but in a time frame quite close to it. The When holds the key, we believe, not only to the Who of Stonehenge I, but also to the Why of Stonehenge II and III.

For, as we shall see, the hurried remaking of Stonehenge in 2100–2000 B.C. had to do with the coming of a New Age—Mankind's first historically recorded New Age.

3

THE TEMPLES
THAT FACED HEAVEN

The more we know about Stonehenge thanks to modern science, the more incredible Stonehenge becomes. Indeed, were it not for the visible evidence of megaliths and earthworks—were they somehow to vanish, as so many ancient monuments did through the vagaries of time and nature or ravages wrought by Man—the whole tale of stones that could compute time and circles that could foretell eclipses and determine the movements of the Sun and the Moon would have sounded so implausible for Stone Age Britain that it would have been considered just a myth.

The great antiquity of Stonehenge, which kept increasing as scientific knowledge about it progressed, is of course what troubles most scientists; and it is primarily the dates of construction ascertained for Stonehenge I and II + III that have led archaeologists to seek Mediterranean visitors, and eminent scholars to allude to ancient gods, as the only possible explanations for the enigma.

For of the series of troubling questions, such as by Whom and What for, the When is the most satisfactorily answered. Archaeology and physics (through modern dating methods such as carbon-14 measurements) were joined by *archaeoastronomy* to agree on the dates: 2900/2800 B.C. for Stonehenge I, 2100/2000 B.C. for Stonehenge II and III.

The father of the science of archaeoastronomy—though he preferred to call it astro-archaeology, which better conveys what he had in mind—was undoubtedly Sir Norman Lockyer. It is a measure of how long established science takes to accept innovation to note that it is virtually a full century since the publication of Lockyer's masterwork, *The*

Dawn of Astronomy, in 1894. Having visited the Levant in 1890 he observed that whereas for the early civilizations in India and China there are few monuments but many written records establishing their age, the opposite holds true in Egypt and Babylonia: they were "two civilizations of undefined antiquity" where monuments abounded but the measure of their antiquity was uncertain (at the time of Lockyer's writing).

It struck him, he wrote, that it was truly remarkable that in Babylonia "from the beginning of things the sign for God was a star" and that likewise in Egypt, in the hieroglyphic texts, three stars represented the plural term "gods." Babylonian records on clay tablets and burned clay bricks, he noted, appeared to deal with regular cycles of "moon and planet positions with extreme accuracy." Planets, stars, and the constellations of the zodiac are represented on the walls of Egyptian tombs and on papyruses. In the Hindu pantheon, he observed, we find the worship of the Sun and of the Dawn: the name of the god Indra meaning "The Day Brought by the Sun" and that of the goddess Ushas meaning "Dawn."

Can astronomy be of assistance to Egyptology? he wondered; can it help define the measure of Egyptian and Babylonian antiquity?

When one considers the Hindu *Rigveda* and Egyptian inscriptions from an *astronomical* point of view, Lockyer wrote, "one is struck by the fact that in both, the early worship and all the early observations related to the horizon . . . This was true not only of the Sun, but equally true of the stars which studded the expanse of the sky." The horizon, he pointed out, is "the place where the circle which bounds our view of Earth's surface and the sky appear to meet." A circle, in other words, where Heaven and Earth touch and meet. It is there that the ancient peoples sought whatever sign or omen their observers were looking for. Since the most regular phenomenon observable on the horizon was the rising and setting of the Sun on a daily basis, it was natural to make this the basis of ancient astronomical observations, and to relate other phenomena (such as the appearance or movements of planets and even stars) to their "heliacal rising," their brief appearance on the eastern horizon as the turning Earth reaches the few moments of dawn, when the Sun begins to rise but the sky is dark enough to see the stars.

An ancient observer could easily determine that the Sun always rises in the eastern skies and sets in the western skies, but he would have noted that in summer the Sun appears to rise in a higher arc than in winter and the days are longer. This, modern astronomy explains, is due to the fact that the Earth's axis around which it rotates daily is not perpendicular to its path around the Sun (the Ecliptic) but is inclined to it—about 23.5 degrees nowadays. This creates the seasons and the four points in the seeming movement of the Sun up and down in the skies: the summer and winter solstices and the spring ("vernal") and autumnal equinoxes (which we have described earlier).

Studying the orientation of temples old and not so old, Lockyer found that those he called "Sun Temples" were of two kinds: those oriented according to the equinoxes and those oriented according to the solstices. Though the Sun always rises in the eastern skies and sets in the western skies, it is only on the days of the equinoxes that it rises anywhere on Earth precisely in the east and sets precisely in the west, and Lockyer therefore deemed such "equinoctial" temples to be more universal than those whose axis was oriented according to the solstices; because the angle formed by the northern and southern (to an observer in the northern hemisphere, the summer and winter) solstices depended on where the observer was—his latitude. Therefore, "solstitial" temples were more individual, specific to their geographic location (and even elevation).

As examples of equinoctial temples Lockyer cited the Temple to Zeus at Baalbek, the Temple of Solomon in Jerusalem, and the great basilica of St. Peter's in the Vatican in Rome (Fig. 13)—all oriented on a precise east–west axis. Regarding the latter he quoted studies on church architecture that described how at the Old St. Peter's (begun under Constantine in the fourth century and torn down early in

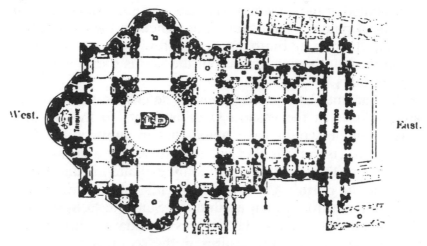

Figure 13

the sixteenth century), on the day of the vernal equinox, "the great doors of the porch of the quadriporticus were opened at sunrise, and also the eastern doors of the church; and as the sun rose, its rays passed through the outer doors, then through the inner doors, and, penetrating straight through the nave, illuminated the High Altar." Lockyer added that "the present church fulfils the same conditions." As examples of "solstitial" Sun Temples Lockyer described the principal Chinese "Temple of Heaven" in Peking, where "the most important of all state observances in China, the sacrifice performed in the open air at the south altar of the Temple of Heaven," was held on the day of the winter solstice, December 21; and the structure at Stonehenge, oriented to the summer solstice.

All that was, however, just a prelude to Lockyer's main studies, in Egypt.

Studying the orientation of Egypt's ancient temples, Lockyer concluded that the older ones were "equinoctial" and the later ones "solstitial." He was amazed to discover that earlier temples revealed greater astronomical sophistication than later ones, for they were intended to observe and venerate not only the rising or setting of the Sun, but also of stars. Moreover, the earliest shrine suggested a mixed

Sun-Moon worship that shifted to an equinoctial, i.e. Sun, focus. That equinoctial shrine, he wrote, was the temple in Heliopolis ("City of the Sun" in Greek) whose Egyptian name, *Annu,* was also mentioned in the Bible as On. Lockyer calculated that the combination of solar observations with the period of the bright star Sirius and with the annual rising of the Nile, a triple conjunction on which the Egyptian calendar was based, indicated that in Egyptian time reckoning Point Zero was circa 3200 B.C.

The Annu shrine, it is known from Egyptian inscriptions, held the *Ben-Ben* ("Pyramidion-Bird"), claimed to have been the actual conical upper part of the "Celestial Barge" in which the god Ra had come to Earth from the "Planet of Millions of Years." This object, usually kept in the temple's inner sanctum, was put on public display once a year, and pilgrimages to the shrine to view and venerate the sacred object continued into dynastic times. The object itself has vanished over the millennia; but a stone replica thereof has been found, showing the great god visible through the doorway or hatch of the capsule (Fig. 14). The legend of the Phoenix, the mythical bird that dies and resurrects after a certain period, has also been traced to this shrine and its veneration.

Figure 14

The *Ben-Ben* was still there at the time of the Pharaoh Pi-Ankhi (circa 750 B.C.), for an inscription was found describing his visit to the shrine. Intent on entering the Holy of Holies and seeing the celestial object, Pi-Ankhi began the process by offering elaborate sacrifices at sunrise in the temple's forecourt. He then entered the temple proper, bowing low to the great god. A prayer was then offered by the priests for the king's safety, that he might enter and leave the Holy of Holies unharmed. Ceremonies that included the king's washing and purification and rubbing with incense then followed, preparing him for entering the enclosure called the "Star Room." He was then given rare flowers or plant branches that had to be offered to the god by placing them in front of the *Ben-Ben*. He then went up the steps to the "great tabernacle" that held the sacred object. Reaching the top of the stairs, he pushed back the bolt and opened the doors to the Holy of Holies; "and he saw his forefather Ra in the chamber of the *Ben-Ben*." He then stepped back, closed the doors behind him and placed thereon a clay seal, impressing on it his signet.

While this shrine has not survived the millennia, what may have been another later shrine modeled after the Heliopolitan one has been found by archaeologists. It is the so-called Solar Temple of the Pharaoh Ne-user-Ra of the fifth dynasty that lasted from 2494 to 2345 B.C. Built at a place now called Abusir, just south of Giza and its great pyramids, it consisted primarily of a large raised terrace upon which, within a great enclosure, there stood on a massive platform a thick, short obelisklike object (Fig. 15). A ramp, surmounted by a covered corridor lighted by evenly spaced windows in the ceiling, connected the temple's elaborate entrance with a monumental gateway in the valley below. The sloping base of the obelisklike object rose some sixty-five feet above the level of the temple's court; the obelisk, which may have been sheathed with gilded copper, rose another 120 feet.

The temple, within its walled enclosure which contained various chambers and compartments, formed a perfect rectangle measuring some 260 feet by 360 feet. It was clearly

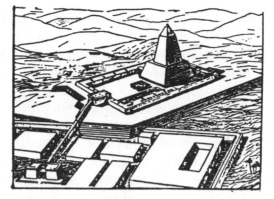

Figure 15

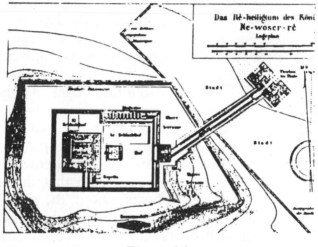

Figure 16

oriented on an east–west axis (Fig. 16), i.e. to the equinoxes; but the long corridor was obviously reoriented away from the east–west axis to face northeast. That this was a deliberate reorientation of a copy of the earlier Heliopolitan shrine (which was oriented just on the east–west axis) becomes clear from the masterful reliefs and inscriptions that decorated the corridor. They celebrated the thirtieth anniversary of the pharaoh's reign, so the corridor could have been built then. The celebration followed the mysterious rites of the *Sed* festival (the term's meaning remains unclear) which marked some kind of "jubilee" and which always began on the first day of the Egyptian calendar—the first day of the first month which was named the Month of Thoth. In other words, the Sed festival was a kind of New Year's Day festival celebrated not each year but after the passage of a number of years.

The presence of both equinoctial and solstitial orientations in this temple implies familiarity—in the third millennium B.C.—with the concept of the Four Corners. Drawings and inscriptions found in the temple's corridor describe the king's "sacred dance." They were copied, translated, and published by Ludwig Borchardt with H. Kees and Friedrich von Bissing in *Das Re-Heiligtum des Königs Ne-Woser-Re*. They concluded that the "dance" represented the "cycle of sanctification of the four corners of the Earth."

The equinoctial orientation of the temple proper and the solstitial one of the corridor, bespeaking the movements of the Sun, led Egyptologists to apply to the structure the term "Sun Temple." They found reinforcement in this designation in the discovery of a "solar boat" (partly carved out of the rock and partly built of dried and painted bricks) buried under the sands just south of the temple enclosure. Hieroglyphic texts dealing with the measurement of time and the calendar in ancient Egypt held that the celestial bodies traversed the skies in boats. Often, the gods or even the deified pharaohs (having joined the gods in the Afterlife) were depicted in such boats, sailing above the firmament of the skies that was held up at the four corner points (Fig. 17).

The next great temple clearly emulated the pyramidion-on-platform concept (Fig. 18) of the Ne-User-Ra "Sun Temple"; but it was already fully oriented to the solstices from its inception, having been planned and executed along a northwest–southeast axis. It was built on the west side of the Nile (near the present-day village of Deir-el-Bahari) in Upper Egypt, as part of greater Thebes, by the Pharaoh Mentuhotep I circa 2100 B.C. Six centuries later Tuthmosis III and Queen Hatshepsut of the XVIII dynasty added their temples there; the orientation was similar—but not exactly so (Fig. 19). It was at Thebes (Karnak) that Lockyer made

Figure 17

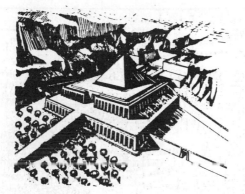

Figure 18

his most important discovery, one that laid the foundation for archaeoastronomy.

The sequence of chapters, facts, and arguments in *The Dawn of Astronomy* reveals that Lockyer's route to Karnak and Egyptian temples passed through the European evidence. There was the orientation of the Old St. Peter's in Rome and the information about the beam of sunlight at the spring equinox sunrise; and there was St. Peter's Square (a woodcut drawing of which Lockyer included, Fig. 20) with its startling similarities to Stonehenge . . .

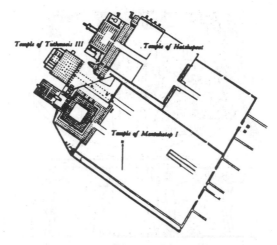

Figure 19

Figure 20

He looked at the Parthenon in Athens, Greece's principal shrine (Fig. 21) and found that "there is the old Parthenon, a building which may have been standing at the time of the Trojan war, and the new Parthenon, with an outer court very like the Egyptian temples but with its sanctuary more nearly in the centre of the building. It was by the difference of direction of these two temples at Athens that my attention was called to the subject."

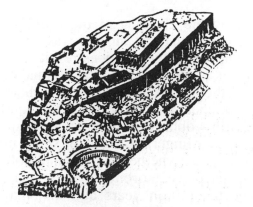

Figure 21

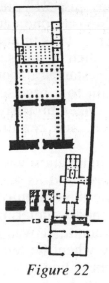

Figure 22

He had in front of him drawings of the layout plans of various Egyptian temples where orientations seemed to vary from early to later buildings, and was struck by an obvious one in two back-to-back temples at a site not far from Thebes called Medinet-Habu (Fig. 22) and pointed out the similarity between this Egyptian and the Greek ''difference of orientation'' in temples that, from a purely architectural aspect, should have been parallel and with the same axial orientation.

Could the slightly altered orientation result from changes in the amplitude (the position in the skies) of the Sun or stars caused by the changes in the Earth's obliquity? he wondered, and felt that the answer was Yes.

We now know that the solstices result from the fact that the Earth's axis is tilted relative to its plane of orbit around the Sun, and the points of "standstill" match the Earth's tilt. But astronomers established that this angle is not constant. The Earth wobbles, like a pitching ship, from side to side—perhaps the lingering result of some mighty bang it received in its past (whether the original collision that put the Earth in its present orbit, or the crash of a massive meteor some 65 million years ago that may have extinguished the dinosaurs). The present tilt of about 23.5 degrees can decrease to perhaps just 21° and on the other hand increase to well over 24°—no one can really say for sure, since the change by even 1° lasts thousands of years (7,000, according to Lockyer). Such changes in the obliquity result in changes in the Sun's standstill points (Fig. 23a). This means that a temple built to a precise solstitial orientation at a given time is no longer properly aligned to that orientation several hundred, and certainly several thousand, years later.

Lockyer's masterful innovation was this: by determining the orientation of a temple and its geographic longitude, it was possible to calculate the obliquity that prevailed at the time of construction; and by determining the changes in obliquity over the millennia, it was possible to conclude with sufficient certainty when the temple was constructed.

The Table of Obliquity, fine-tuned and made more accurate during the past century, shows the change in the angle of the Earth's tilt in five-hundred-year intervals, going back from the present 23° 27' (about 23.5 degrees):

500 B.C.	about	23.75 degrees	
1000 B.C.	"	23.81	"
1500 B.C.	"	23.87	"
2000 B.C.	"	23.92	"
2500 B.C.	"	23.97	"
3000 B.C.	"	24.02	"
3500 B.C.	"	24.07	"
4000 B.C.	"	24.11	"

Lockyer applied his findings primarily to extensive measurements at the great temple to Amon-Ra in Karnak. This temple, having been enlarged and augmented by various pharaohs, consists of two principal rectangular structures built back-to-back on a southeast–northwest axis, signifying a solstitial orientation. Lockyer concluded that the purpose of the orientation and the layout of the temple was to enable a beam of sunlight to come from such a direction on solstice day that it would travel the length of a long corridor, pass between two obelisks, and strike the Holy of Holies with a flash of Divine Light at the temple's innermost sanctum. Lockyer noticed that the axis of the two back-to-back temples was not similarly oriented: the newer axis represented a solstice resulting from a somewhat smaller obliquity than the older axis (Fig. 23b). The two obliquities determined by Lockyer show that the older temple was built circa 2100 B.C. and the newer one circa 1200 B.C.

Although more recent investigations, especially by Gerald S. Hawkins, suggest that the Sun's beam, at winter solstice, was meant to be viewed from a part of the temples Hawkins named "High Room of the Sun" and not as a beam traveling the length of the axis, the revision in no way changes the basic conclusion by Locker regarding the solstitial orien-

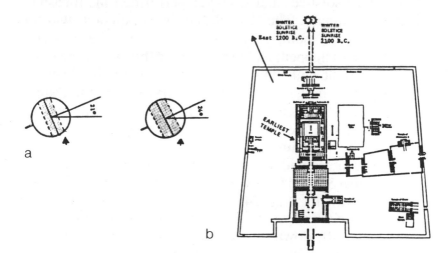

Figures 23a and 23b

tation. Indeed, further archaeological discoveries at Karnak corroborate Lockyer's principal innovation—that the orientation of the temples changed in time to reflect the changes in obliquity. Therefore, the orientation could serve as a clue to the temples' time of construction. The latest archaeological advances confirmed that the construction of the oldest part coincided with the beginning of the Middle Kingdom under the XI dynasty circa 2100 B.C. Repairs, demolitions, and rebuilding then continued through the ensuing centuries by pharaohs of subsequent dynasties; the two obelisks were set up by pharaohs of the XVIII dynasty. The final phase took shape under the Pharaoh Seti II of the XIX dynasty who reigned in 1216–1210 B.C.—all as Lockyer had determined.

Archaeoastronomy—or, astro-archaeology as Sir Norman Lockyer named it—proved its merit and validity.

At the beginning of this century Lockyer turned his attention to Stonehenge, having become convinced that the phenomenon he had discovered governed temple orientations in other parts of the ancient world, as at the Parthenon in Athens. At Stonehenge the axis of viewing from the center through the Sarsen Circle clearly bespoke an orientation to the summer solstice, and Lockyer performed his measurements accordingly. The Heel Stone, he concluded, was the indicator of the point on the horizon where the expected sunrise was to happen; and the apparent shifting of the stone (with attendant widening and realignment of the Avenue) suggested to him that as the centuries passed and the change in the Earth's tilt kept changing the sunrise point, even if ever so slightly, the people in charge of Stonehenge kept adjusting the view line.

Lockyer presented his conclusions in *Stonehenge and Other British Stone Monuments* (1906); they can be summed up in one drawing (Fig. 24). It assumes an axis that begins at the Altar Stone, passes between the sarsen stones numbered 1 and 30, down the Avenue, toward the Heel Stone as the focusing pillar. The obliquity angle indicated by such an axis led him to suggest that Stonehenge was built in 1680

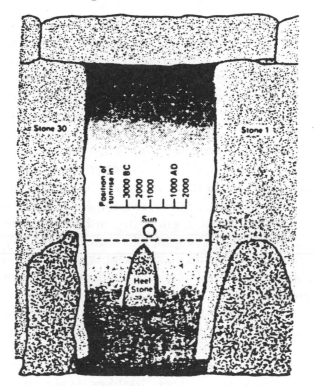

Figure 24

B.C. Needless to say, such an early date was quite sensational at a time, a century ago, when scholars still thought of Stonehenge in terms of King Arthur's days.

The refinements in studies of the Earth's obliquity, allowances now made for margins of error, and the determination of the various phases of Stonehenge have not diminished Lockyer's basic contribution. Although Stonehenge III, which is what we essentially see nowadays, is now dated to circa 2000 B.C., it is generally agreed that the Altar Stone was removed when the remodeling began circa 2100 B.C. with the double Bluestone Circle (Stonehenge II), and that it was reerected where it is now only when the bluestones were reintroduced and the Y and Z holes dug. That phase, designated Stonehenge IIIb, has not been definitely dated; it is in a range between 2000 B.C. (Stonehenge IIIa) and 1550 B.C. (Stonehenge IIIc)—and quite possibly the 1680 B.C. date arrived at by Lockyer. As the drawing shows, he did not rule out a much earlier date for the prior phases of Stonehenge; this too compares well with the presently accepted date of 2900/2800 B.C. for Stonehenge I.

Archaeoastronomy thus joins archaeological findings and radiocarbon dating to arrive at the same dates for the construction of the various phases of Stonehenge, the three separate methods corroborating each other. With such a convincing determination of Stonehenge's dates, the question regarding its builders becomes more poignant. Who, circa 2900/2800 B.C., possessed the knowledge of astronomy (to say nothing of engineering and architecture) to build such a calendrical "computer," and circa 2100/2000 B.C. to rearrange the various components thereof and attain a new realignment? And why was such a realignment required or desired?

Mankind's transition from the Paleolithic (Old Stone Age) that lasted for hundreds of thousands of years to the Mesolithic (Middle Stone Age) occurred abruptly in the ancient Near East. There, circa 11,000 B.C.—right after the Deluge, according to our calculations—deliberate agriculture and the domestication of animals began in a stunning profusion. Archaeological and other evidence (most recently augmented by studies of linguistic patterns) shows that Mesolithic agriculture spread from the Near East to Europe as a result of the migration of people possessing such knowledge. It reached the Iberian peninsula between 4500 and 4000 B.C., the western edge of what is today France and the Lowlands between 3500 and 3000 B.C., and the British Isles between 3000 and 2500 B.C. It was soon thereafter that the "Beaker People," who knew how to make clay utensils, arrived on the Stonehenge scene.

But by then the ancient Near East was already well past the Neolithic (New Stone Age) which began there circa 7400 B.C. and whose hallmark was the transition from stone to clay to metals and the appearance of urban settlements. By the time this phase reached the British Isles with the so-called "Wessex People" (after 2000 B.C.), in the Near East the great Sumerian civilization was already almost two thousand years old and the Egyptian civilization more than a thousand years old.

If, as all agree, the sophisticated scientific knowledge that was required for the planning, siting, orientation, and construction of Stonehenge had to come from outside the British Isles, the earlier civilizations of the Near East seem to be the only sources for such knowledge at the time.

Were the Sun Temples of Egypt, then, the prototypes for Stonehenge? We have seen that at the dates established for Stonehenge's various phases, there already existed in Egypt elaborate temples that were astronomically oriented. The equinoctial Sun Temple at Heliopolis was built at about the time, 3100 B.C., when kingship began in Egypt (if not somewhat earlier)—several centuries before Stonehenge I. The construction of the oldest phase of the solstitially oriented temple to Amon-Ra in Karnak took place circa 2100 B.C.—a date coinciding (perhaps not by chance) with the date for the "remodeling" of Stonehenge.

It is thus theoretically possible that Mediterranean people—Egyptians or people with "Egyptian" knowledge—could somehow account for the construction of Stonehenge I, II, and III at dates that were impossible for the local inhabitants of the area.

While, from a timing point of view, Egypt could have been the source of the required knowledge, we ought to be bothered by a crucial difference between *all* of the Egyptian temples and Stonehenge: none of the Egyptian temples, no matter whether their orientation was equinoctial or solstitial, were circular as Stonehenge has been during all of its phases. The various pyramids were square-based; the podiums for the obelisks and pyramidions were square; the numerous temples were all rectangular. With all the stones of Egypt, not one of its temples was a stone *henge*.

From the beginning of dynastic times in Egypt, with which the appearance of a distinct Egyptian civilization is linked, it was the pharaohs of Egypt who had hired the architects and masons, the priests and savants, and decreed the planning and construction of the marvelous stone edifices of ancient Egypt. None of them, however, appears to have designed, oriented, and built a circular temple.

What about those famous seafarers, the Phoenicians? Not only did they reach the British Isles (mainly in search of tin) too late to have built not just Stonehenge I but also the II and III phases, but none of their temple architecture bears any resemblance to the emphatically circular essence of Stonehenge. We can see a Phoenician temple depicted on the Byblos coin (Fig. 12), and it is certainly rectangular. On the vast stone platform at Baalbek in the Lebanon mountains, people after people and conquerors after conquerors built their temples precisely on the ruins and according to the layout of preceding temples. These, as the latest extant ruins from the Roman era reveal (Fig. 25), represented a rectangular temple (black area) with a square forecourt (the diamond-shaped entrance pavilion is a purely Roman addition). The temple is clearly oriented on an east–west axis, facing directly east toward the Sun at sunrise—an equinoctial temple. This should perhaps be no surprise, since in ancient times this site too was called "City of the Sun"— Heliopolis by the Greeks, Beth-Shemesh ("House of the Sun") in the Bible, in King Solomon's time.

That the rectangular shape and east–west axis were not a passing fad in Phoenicia is further evidenced by the Temple of Solomon, the first temple of Jerusalem, which was built with the aid of Phoenician architects provided by

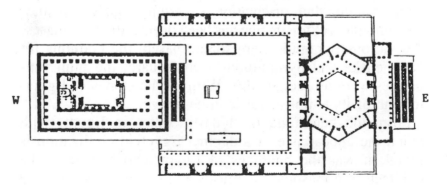

Figure 25

Ahiram, king of Tyre; it was a rectangular structure on an east–west axis, facing eastward (Fig. 26), built upon a large man-made platform. Sabatino Moscati (*The World of the Phoenicians*) stated without qualification that "if there are no adequate remains of Phoenician temples, the temple of Solomon in Jerusalem, built by Phoenician workmen, *is* described in detail in the Old Testament—and the Phoenician temples must have resembled each other." And nothing about them was circular.

Circles do appear, though, in the case of the other Mediterranean "suspects"—the Mycenaeans, the first Hellenic people of ancient Greece. But these were at first what archaeologists call Grave Circles—burial pits surrounded by a circle of stones (Fig. 27) that evolved into circular tombs hidden beneath a conical mound of soil. But that had taken place circa 1500 B.C. and the largest of them, called the Treasury of Atreus because of the golden artifacts that were found around the dead (Fig. 28), dates to circa 1300 B.C. Archaeologists who adhere to the Mycenaean connection compare such eastern Mediterranean burial mounds to Silbury Hill in the Stonehenge area or to one at Newgrange, across the Irish Sea in Boyne Valley, County Meath, in Ireland; but Silbury Hill has been determined by carbon

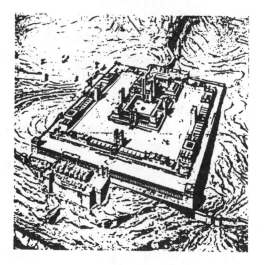

Figure 26

Figure 27

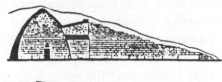

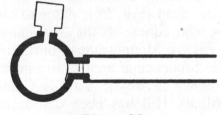

Figure 28

dating to have been constructed not later than 2200 B.C. and the burial mound at Newgrange at about the same time—almost a thousand years before the Treasury of Atreus and other Mycenaean examples; the period of the Mycenean burial mounds, moreover, is even farther removed from the time of Stonehenge I. In fact, the burial mounds in the British Isles are much more akin, in construction and in timing, to such mounds in the western rather than eastern Mediterranean, such as the one in Los Millares in southern Spain (Fig. 29).

Above all, Stonehenge has never served as a burial place. For all these reasons, the search for a prototype—a circular structure serving astronomical purposes—should continue beyond the eastern Mediterranean.

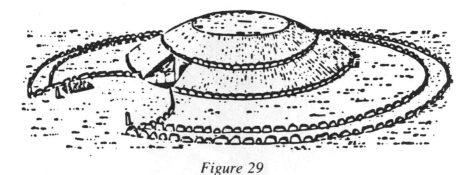

Figure 29

* * *

Older than the Egyptian civilization and possessing much more advanced scientific knowledge, the Sumerian civilization could have served, theoretically, as the fountainhead for Stonehenge. Among the astounding Sumerian achievements were great cities, a written language, literature, schools, kings, courts, laws, judges, merchants, craftsmen, poets, dancers. The sciences flourished within the temples, where the "secrets of numbers and of the heavens"—of mathematics and astronomy—were kept, taught, and transmitted by generations of priests who performed their functions within walled-off sacred compounds. Such compounds usually included shrines dedicated to various deities, residences, work and study places for the priests, storehouses and other administrative buildings, and—as the dominant, principal, and most prominent feature of the sacred precinct and of the city itself—a *ziggurat*, a pyramid that rose sky high in stages (usually seven). The topmost stage was a multichambered structure that was intended—literally—to be the residence of the great god whose "cult center" (as scholars like to call it) the city was (Fig. 30).

A good illustration of the layout of such a sacred precinct with its ziggurat is a reconstruction based on archaeological discoveries at the sacred precinct of Nippur (NI.IBRU in Sumerian), the "headquarters" from the earliest days of the god Enlil (Fig. 31); it shows a ziggurat with a square base within a rectangular compound. As luck would have it, archaeologists have also unearthed a clay tablet upon which

Figure 30

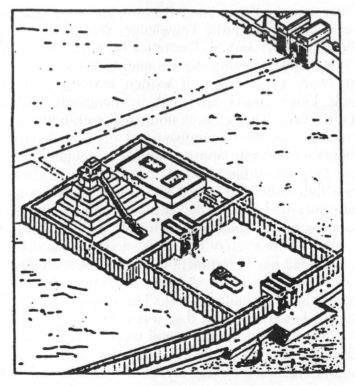

Figure 31

an ancient cartographer drew a map of Nippur (Fig. 32); it
clearly shows the rectangular sacred precinct with the
square-based ziggurat, the caption (in cuneiform script) for
which states its name, the E.KUR—''House, which is like

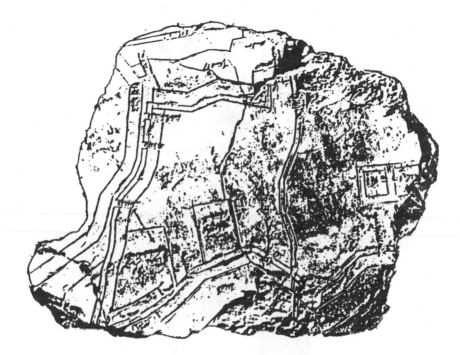

Figure 32

a mountain.'' The orientation of the ziggurat and temples
was such that the corners of the structures pointed to the
four cardinal points of the compass, so that the sides of the
structure faced to the northeast, southwest, northwest, and
southeast.

To orient the ziggurats' corners to the cardinal points—
without a compass—was not an easy achievement. But it
was an orientation that made it possible to scan the heavens
in many directions and angles. Each stage of the ziggurat
provided a higher viewing point and thus a different horizon,
adjustable to the geographic location; the line between the
east-pointing and west-pointing corners provided the equinoctial orientation; the sides gave solstitial views to either
sunrise or sunset, at both summer and winter solstices. Modern astronomers have found many of these observational
orientations in the famed ziggurat of Babylon (Fig. 33)

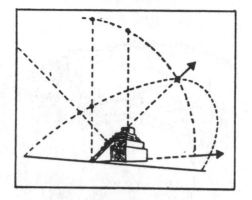

Figure 33

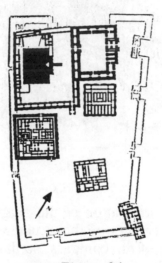

Figure 34

whose precise measurements and building plans were found
spelled out on clay tablets.

Square or rectangular structures, with precise right angles, were the traditional shape of Mesopotamian ziggurats
and temples, whether one looks at the sacred precinct of Ur
at the time of Abraham (Fig. 34)—circa 2100 B.C., the time
of Stonehenge II—or goes back to one of the earliest temples
built on a raised platform, as the White Temple at Eridu
(Fig. 35a and 35b) that dates to about 3100 B.C.—two or
three centuries before the date of Stonehenge I.

The deliberate manner in which the Mesopotamian tem-

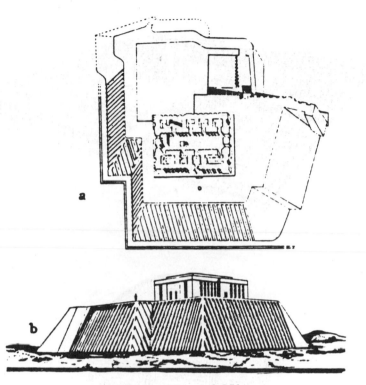

Figures 35a and 35b

plcs, at all times, were given the rectangular shape and
specific orientation can be easily inferred from the layout
in Babylon by comparing the haphazard meshing of buildings and alleys in cities in Babylonian times with the straight
and geometrically perfect layout of the sacred precinct of
Babylon and the square shape of its ziggurat (Fig. 36).

Mesopotamian temples were thus rectangular and ziggurats square-based quite deliberately. In case someone
wonders whether this was because the Sumerians and their
successors were unfamiliar with the circle or unable to construct one, suffice it to point out that in mathematical tablets
certain key numbers of the sexagesimal (''base 60'') system
were represented by circles; in tablets dealing with geometry
and land measurement, instructions were given for measuring regular- and irregular-shaped areas, including circles.
The round wheel was known (Fig. 37)—another Sumerian

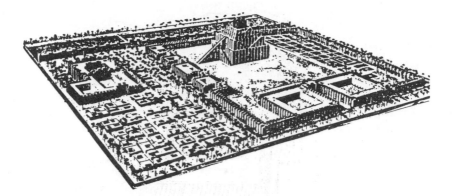

Figure 36

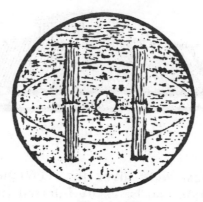

Figure 37

"first." Obviously circular residential houses were found in the ruins of early cities (Fig. 38); a sacred precinct (as this one at a site called Khafajeh—Fig. 39) was sometimes surrounded by an oval-forming wall. It is clear that avoiding a well-known circular shape for temples was deliberate.

There were thus basic design, architecture, and orientation differences between Sumerian temples and Stonehenge, to which one could add the fact that the Sumerians were not stonemasons (there being no stone quarries in the alluvial plain between the Euphrates and Tigris rivers). The Sumerians were not the ones who planned and erected Stone-

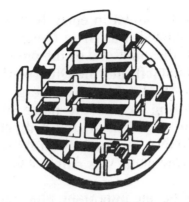

Figure 38

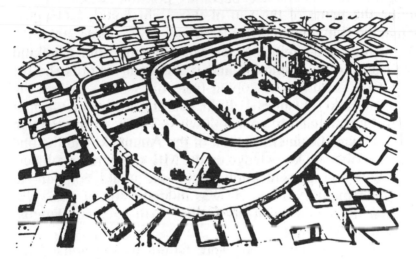

Figure 39

henge; and the only instance that can be considered an exception to discoveries and Sumerian temples, as we shall see, reinforces this conclusion.

So, if not the Egyptians or Phoenicians or early Greeks, if not the Sumerians and their successors in Mesopotamia—who then came to the plain of Salisbury to plan and supervise the erection of Stonehenge?

An interesting clue emerges as one reads the legends concerning the tumulus of Newgrange. According to Michael J. O'Kelly, a leading architect and explorer of the site and its surroundings (*Newgrange: Archaeology, Art and*

Legend), the site was known in early Irish lore by various names that all designated it as *Brug Oengusa,* the "House of Oengus," son of the chief god of the pre-Celtic pantheon who had come to Ireland from "the Otherworld." That chief god was known as *An Dagda,* "An, the good god" . . .

It is indeed amazing to find the name of the principal deity of the ancient world in all these diverse places—in Sumer and his E.ANNA ziggurat of Uruk; in the Egyptian Heliopolis, whose true name was Annu; and in far-removed Ireland . . .

That this might be an important clue and not just an insignificant coincidence becomes possible when we examine the name of the son of this "chief god," Oengus. When the Babylonian priest Berossus wrote, circa 290 B.C., the history and prehistory of Mesopotamia and Mankind according to the Sumerian and Babylonian records, he (or the Greek savants who copied from his works) spelled the name of Enki "Oannes." Enki was the leader of the first group of Anunnaki to splash down to Earth, in the Persian Gulf; he was the chief scientist of the Anunnaki and the one who inscribed all knowledge on the ME's, enigmatic objects that, with our present knowledge, one could compare to computer memory discs. He was indeed a son of Anu; was he then the god who in pre-Celtic myth became Oengus, the son of An Daga?

"All that we know, we were taught by the gods," the Sumerians repeatedly stated.

Was it, then, not the ancient *peoples,* but the ancient *gods* who created Stonehenge?

4

DUR.AN.KI—THE "BOND HEAVEN-EARTH"

From the earliest days, Man has lifted his eyes to the heavens for divine guidance, for inspiration, for help in troubled times. From the very beginning, even as Earth was separated from "Heaven" when it was created, heaven and Earth continued to meet everlastingly on the horizon. It was there, as Man gazed into the distance, at sunrise or sunset, that he could see the Heavenly Host.

Heaven and Earth meet on the horizon, and the knowledge based on observing the skies and the celestial motions resulting therefrom is called Astronomy.

From the earliest days, Man knew that his creators had come from the heavens—*Anunnaki* he called them, literally "Those who from Heaven to Earth Came." Their true abode was in the heavens, Man always knew: "Father who art in Heaven," Man knew to say. But those of the Anunnaki who had come and stayed on Earth, Man also knew, could be worshiped in the temples.

Man and his gods met in the temples, and the knowledge and ritual and beliefs that resulted are called Religion.

The most important "cult center," the "navel of the earth," was Enlil's city in what was later Sumer. Central religiously, philosophically, and actually, that city, Nippur, was the Mission Control Center; and its Holy of Holies, where the Tablets of Destinies were kept, was called DUR.AN.KI—"Bond Heaven-Earth."

And ever since, at all times and in all places and in all religions, the places of worship that are called temples, in spite of all the changes that they, and Mankind and its

79

religions have undergone, have remained the *Bond Heaven-Earth*.

In ancient times astronomy and religion were linked: the priests were the astronomers and astronomers were priests. When Yahweh made his covenant with Abraham, He instructed Abraham to step out and lift his gaze skyward to try and count the stars. There was more than an idle stratagem in this, for Abraham's father, Terah, was an oracle priest in Nippur and Ur and thus knowing in astronomy.

In those days each of the Great Anunnaki was assigned a celestial counterpart, and since the Solar System had twelve members, the "Olympic Circle," throughout the millennia and up to and including Greek time, was always made up of twelve. It was thus that the worship of the gods was closely associated with the motions of the celestial bodies, and the biblical admonitions against the worship of "the Sun, the Moon and the Host of Heaven" were in reality admonitions against the worship of gods other than Yahweh.

The rituals, festivals, days of abstinence, and other rites that expressed the worship of the gods were thus attuned to the motions of the gods' celestial counterparts. Worship required a calendar; temples were observatories; priests were astronomers. The ziggurats were Temples of Time, where time-keeping joined astronomy to formalize worship.

And Adam knew his wife again
and she bore a son and called his name Sheth,
for God (she said) has granted me another offspring
instead of Abel, whom Cain slew.
And to Sheth, in turn, a son was born
and he called his name Enosh.
It was then that calling Yahweh by name began.

Thus, according to the Bible (Genesis 4:25–26), did the Children of Adam begin to worship their God. How this calling in the name of the Lord was done—what form the worship took, what rituals were involved—we are not told. It happened, the Bible makes clear, in remote times, well before the Deluge. Sumerian texts, however, throw light on the subject. They not only assert—repeatedly and emphatically—that there were Cities of the Gods in Mesopotamia before the Deluge, and that when the Deluge had occurred there had already been "demigods" (offspring of "Daughters of Man" by male Anunnaki "gods"), but also that the worship took place in consecrated places (we call them "temples"). They were already, we learn from the earliest texts, Temples of Time.

One of the Mesopotamian versions of the events leading to the Deluge is the text known (by its opening words) *"When the gods like men"* in which the hero of the Deluge is called *Atra Hasis* ("He who is exceedingly wise"). The tale relates how Anu, the ruler of Nibiru, returned to that planet from a visit to Earth after arranging a division of powers and territories on Earth between his feuding sons, the half brothers Enlil ("Lord of the Command") and Enki ("Lord of Earth"), putting Enki in charge of the gold-mining operations in Africa. After describing the hard work of the Anunnaki assigned to the mines, their mutiny, and the ensuing creation through genetic engineering by Enki and his half sister Ninharsag of the *Adamu,* a "Primitive Worker," the epic relates how Mankind began to procreate and multiply. In time, Mankind began to upset Enlil by its excessive "conjugations," especially with the Anunnaki (a situation reflected in the biblical version of the Deluge tale); and Enlil prevailed on the Great Anunnaki, in their Council, to use the foreseen catastrophe of the avalanche of water to wipe Mankind off the face of the Earth.

But Enki, though he joined in swearing to keep the decision a secret from Mankind, was not happy with the decision and sought ways to frustrate it. He chose to achieve that through the intermediary of Atra-Hasis, a son of Enki by a human mother. The text, which at times assumes a biographical style by Atra-Hasis himself, quotes him saying, "I am Atra-Hasis; I lived in the temple of Enki my lord"—a statement which clearly establishes the existence of a temple in those remote pre-Diluvial times.

Describing the worsening climatic conditions on the one hand and Enlil's harsh measures against Mankind on the other hand in the period preceding the Deluge, the text quotes·Enki's advice to the people through Atra-Hasis how to protest against Enlil's decrees: the worship of the gods should stop!

"Enki opened his mouth and addressed his servant," saying thus to him:

> *The elders, on a sign,*
> *summon to the House of Council.*
> *Let heralds proclaim a command*
> *loudly throughout the land:*
> *Do not reverence your gods,*
> *do not pray to your goddesses.*

As the situation got worse and the catastrophe day neared, Atra-Hasis persisted in his intercession with his god Enki. "In the temple of his god . . . he set foot . . . every day he wept, bringing oblations in the morning." Seeking Enki's help to avert Mankind's demise, "he called by the name of his god"—words that employ the same terminology as in the above-quoted verse from the Bible. In the end Enki decided to subvert the decision of the Council of the Anunnaki by summoning Atra-Hasis to the temple and speaking to him from behind a screen. The event was commemorated on a Sumerian cylinder seal, showing Enki (as the Serpent God) revealing the secret of the Deluge to Atra-Hasis (Fig. 40). Giving him instructions for the building of a submersible boat that would withstand the avalanche of water, Enki advised Atra-Hasis to lose no time, for he had only seven days left before the catastrophe happened. To make sure Atra-Hasis wasted no time, Enki put into motion a clocklike device:

> *He opened the water clock*
> *and filled it;*
> *the coming of the flood on the seventh night*
> *he marked off for him.*

This little-noticed bit of information reveals that time was kept in the temples and that time-keeping goes back to the

Figure 40

earliest, even pre-Diluvial times. It has been assumed that the ancient illustration depicts (on the right) the reed screen from behind which Enki had spoken to the hero of the great flood, the biblical Noah. One must wonder, however, whether what we see is not a reed screen, *but a depiction of that prehistoric water clock* (held up by its priestly attendant).

Enki was the chief scientist of the Anunnaki; it is no wonder, therefore, that it was at his temple, at his "cult center" Eridu, that the first human scientists, the Wise Men, served as priests. One of the first, if not the very first, was called Adapa. Though the original Sumerian Adapa text has not been found, Akkadian and Assyrian versions on clay fragments that have been found attest the tale's significance. Informing us at the very beginning that Adapa's command of wisdom was almost as good as that of Enki himself, the text proceeds to explain that Enki had "perfected for him wide understanding, disclosing all the designs of the Earth; Wisdom he had given to him." It was all done at the temple; Adapa, we are told, "daily did attend the sanctuary of Eridu."

According to Sumerian chronicles of the earlier times, it was at Eridu's temple that Enki, as guardian of the secrets of all scientific knowledge, kept the ME's—tabletlike objects on which the scientific data were inscribed. One of the Sumerian texts details how the goddess Inanna (later known as Ishtar), wishing to give status to her "cult center" Uruk (the biblical Erech), tricked Enki into giving her some of these divine formulas. Adapa, we find, was also nicknamed NUN.ME, meaning "He who can decipher the ME's." Even unto millennia later, in Assyrian times, the saying "Wise as Adapa" meant that someone was exceedingly wise and knowledgeable. The study of sciences was often referred to in Mesopotamian texts as *Shunnat apkali Adapa*, "recital/repetition of the great forefather Adapa." A letter by the Assyrian king Ashurbanipal mentioned that his grandfather, King Sennacherib, was given great knowledge when Adapa had appeared to him in a dream. The "wide knowledge" imparted by Enki to Adapa included writing, medicine, and—according to the astronomical series of tablets UD.SAR.ANUM.ENLILLA ("The Great Days of Anu and Enlil")—knowledge of astronomy and astrology.

Though Adapa had daily attended the sanctuary of Enki, it appears from Sumerian texts that the first officially appointed priest—a function that then passed hereditarily from father to son—was named EN.ME.DUR.AN.KI—"Priest of the ME's of Duranki," the sacred precinct of Nippur. The texts report how the gods "showed him how to observe oil and water, the secrets of Anu, Enlil, and Enki. They gave him the Divine Tablet, the engraved secrets of Heaven and Earth. They taught him how to make calculations with numbers"—the knowledge of mathematics and astronomy, and of the art of measurement, including that of time.

Many of the Mesopotamian tablets dealing with mathematics, astronomy, and the calendar have astounded scientists by their sophistication. At the core of these sciences was a mathematical system called sexagesimal ("Base Sixty") whose advanced nature, including its celestial aspects, has already been discussed. Such sophistication existed even in the earliest times that some call predynastic: arithmetically inscribed tablets (Fig. 41) that have been found attest the use of the sexagesimal system and of numerical record keeping. Designs on clay objects also from the earliest times (Fig. 42) leave no doubt regarding the high level of knowledge of geometry in those remote times,

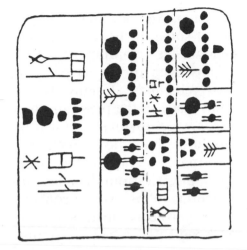

Figure 41

Figure 42

six thousand years ago. And one must wonder whether these designs, or at least some of them, were purely decorative or represented knowledge regarding the Earth, its four "corners," and perhaps even of the shape of astronomically related structures. What these designs also show applies to an important point made in the previous chapter: the circle and circular shapes were obviously known in ancient Mesopotamia and could be drawn to perfection.

Additional information regarding the antiquity of the exact sciences can be gleaned from the tales about Etana, one of the earliest Sumerian rulers. At first considered a mythical hero, he is now recognized as a historical person. According to the Sumerian King Lists, when kingship—an organized civilization—was "lowered again from heaven" after the Deluge, "kingship was first in Kish"—a city whose remains and antiquity have been found and confirmed by archaeologists. Its thirteenth ruler was called Etana, and the King Lists, which by and large only list the names of successive rulers and the length of their reigns, make an exception in the case of Etana by adding after his name the following notation: "A shepherd; he who ascended to heaven, who consolidated all the lands." According to Thorkild Jacobsen (*The Sumerian King List*) Etana's reign began circa 3100 B.C.; excavations at Kish have unearthed the remains of monumental buildings and a ziggurat (stage-temple) dating to the same time.

In the aftermath of the Deluge, when the plain between the Tigris and Euphrates rivers dried sufficiently to enable resettlement, the Cities of the Gods were rebuilt exactly where they had been, according to the "olden plan." Kish, the first City of Men, was entirely new and its place and layout had to be determined. These decisions, we read in the *Tale of Etana*, were made by the gods. Employing scientific knowledge of geometry for layout and astronomy for orientation,

> *The gods traced out a city;*
> *Seven gods laid its foundations.*
> *The city of Kish they traced out,*

and there the seven gods laid its foundation.
A city they established, a dwelling place;
but a Shepherd they withheld.

The twelve rulers at Kish who had preceded Etana were not yet given the Sumerian royal-priestly title EN.SI— "Lordly Shepherd" or as some prefer "Righteous Shepherd." The city, it appears, could attain this status only when the gods could find the right man to build a ziggurat stage-temple there and, by becoming a king-priest, be given the title EN.SI. Who would be "their builder, the one to build the E.HURSAG.KALAMMA," the gods asked— build the "House" (ziggurat) that shall be "Mountainhead for all the lands"?

The task to "look for a king in all the lands, above and below," was assigned to Inanna/Ishtar. She found and recommended Etana—a humble shepherd . . . Enlil, "he who grants kingship," had to make the actual appointment. We read that "Enlil inspected Etana, the young man whom Ishtar had nominated. 'She sought and she found!' he cried. 'In the land shall kingship be established; let the heart of Kish be glad!'"

Now comes the "mythological" part. The brief notation in the King Lists that Etana ascended to heaven stemmed from a chronicle that scholars call the "legend" of Etana which related how Etana, with the permission of the god Utu/Shamash who was in charge of the spaceport, was carried aloft by an "eagle." The higher he rose, the smaller the Earth looked. After the first *beru* of flight the land "became a mere hill"; after the second *beru* the land looked like a mere furrow; after the third *beru*, as a garden ditch; and after one more *beru* the Earth completely disappeared. "As I glanced around," Etana later reported, "the land had disappeared, and upon the sea mine eyes could not feast."

A *beru* in Sumer was a unit of measurement—of length (a "league") and of time (one "double-hour," the twelfth part of a daytime–nighttime period that we now divide into twenty-four hours). It remained a unit of measure in astronomy, when it denoted the twelfth part of the heavenly circle. The text of the *Tale of Etana* does not make clear which unit of measurement—distance, time, or arc—was meant; perhaps all of them. What the text does make clear is that at that remote time, when the first true Shepherd King was enthroned in the first City of Men, distance, time, and the heavens could already be measured.

Kish as the first royal city—under the patronage of "Nimrod"—is mentioned in the Bible (Genesis chapter 10); and certain other aspects of events recorded in the Bible merit exploration. This is especially so because of the puzzling mention in the *Tale of Etana* of the seven gods involved in the planning—and thus orientation—of the city and its ziggurat.

Since all the major gods of ancient Mesopotamia had celestial counterparts from among the twelve members of the Solar System, as well as a counterpart from the twelve constellations of the zodiac and from the twelve months, one must wonder whether the reference to the determination of the orientation of Kish and its ziggurat by the "seven gods" did not actually mean by the seven planets which those deities represented. Were the Anunnaki waiting for the propitious alignment of seven planets as the right time and right orientation for Kish and its ziggurat?

Further light, we believe, can be shed on the subject by journeying in time over more than two thousand years to Judea circa 1000 B.C. Incredibly, we find that about three thousand years ago the circumstances surrounding the selection of a shepherd to be the builder of a new temple in a new royal capital emulated the events and circumstances recorded in the *Tale of Etana;* and the same number seven, with a calendrical significance, also played a role.

The Judean city where the ancient drama was reenacted was Jerusalem. David, who was shepherding the flocks of his father, Jesse the Bethlehemite, was chosen by the Lord for kingship. After the death of King Saul, when David reigned in Hebron over the tribe of Judah alone, representatives of the other eleven tribes "came unto David in Hebron" and asking him to become king over all of them reminded him that Yahweh had earlier said thus to him: "You shall shepherd my people Israel and shall be a *Nagid* over Israel" (II Samuel 5:2).

The term *Nagid* is usually translated "Captain" (King James Version), "Commander" (*The New American Bible*) or even "Prince" (*The New English Bible*). None appear to have realized that *Nagid* is a Sumerian loanword, a term borrowed intact from the Sumerian language, in which the word meant "herdsman"!

A principal preoccupation of the Israelites at that time was the need to find a home for the Ark of the Covenant— not just a permanent home, but also a safe one. Originally made and placed by Moses in the Tent of Appointment during the Exodus, it contained the two stone tablets inscribed with the Ten Commandments on Mount Sinai. Made of specific wood and overlaid with gold both inside and outside, it was surmounted by two Cherubim made of hardened gold with wings extended toward each other; and each time Moses had an appointment with the Lord, Yahweh spoke to him "from between the two Cherubim" (Fig. 43a is a reconstruction suggested by Hugo Gressmann (*Die Lade Jahves*) because of similar depictions found in northern Phoenicia; Fig. 43b is a depiction suggested by A. Parrot in *Le Temple de Jérusalem*). We believe that the Ark, with its insulated gold layers and Cherubim was a communication device, perhaps electrically powered (when it was once touched inadvertently, the person involved fell dead).

Yahweh had given very detailed instructions regarding the construction of the Tent of Appointment and the enclosure for it, and for the Ark, including what amounted to an "operating manual" for the dismantling and reassembly of all that as well as for the careful transportation of the Ark. By David's time, however, the Ark was no longer carried by wooden staves but transported upon a wheeled carriage. It was moved from one temporary place of worship to another, and a major assignment for the newly anointed Shepherd King was to establish a new national capital in Jerusalem and therein build a permanent housing for the Ark in the "House of the Lord."

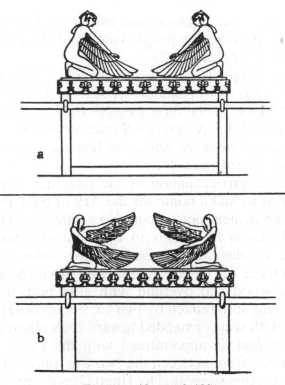

Figures 43a and 43b

But this was not to come to pass. Speaking to King David through the Prophet Nathan, the Lord informed him that it would not be he but his son who would be granted the privilege of building a House of Cedars for Yahweh. And so it was that one of the very first tasks of King Solomon was to build the "House of Yahweh" (now referred to as the First Temple) in Jerusalem. Built as the sacred compound and its components in the Sinai were, it was erected in accordance with very detailed instructions. In fact, the layout plans of the two are almost identical (Fig. 44a the sacred compound in the Sinai; Fig. 44b the Temple of Solomon). And both were oriented along a precise east–west axis, identifying them both as equinoctial temples.

The similarities between Kish and Jerusalem as new national capitals, a Shepherd King, and the task of building a temple whose plans were provided by the Lord is en-

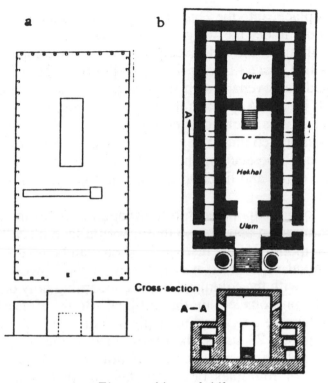

Figures 44a and 44b

hanced by the significance of the number seven.

We are informed in I Kings (chapter 3) that Solomon proceeded to organize the construction project (it involved, among others in the workforce, 80,000 stone quarriers and 70,000 porters) only after Yahweh had appeared unto Solomon in Gibeon "in a nightly vision." The construction, lasting seven years, began with laying the foundation stone in the fourth year of Solomon's reign and "in the eleventh year, in the month of Bul which is the eighth month the Temple was completed in all its stipulations and exactly according to its plans." But although entirely complete with no detail missed or omitted, the Temple was not inaugurated.

It was only eleven months later, "in the month of Etanim, the seventh month, on the festival," that all the elders and tribal chiefs from all over assembled in Jerusalem, "and the priests brought the Ark of the Covenant with Yahweh

into its place, into the *Dvir* of the temple which is the Holy
of Holies, under the wings of the Cherubim . . . And there
was nothing in the Ark except the two stone tablets which
Moses had placed therein in the Wilderness after Yahweh
had made a covenant with the Children of Israel after they
had left Egypt. And when the priests had stepped out of the
Holy of Holies, a cloud filled the House of Yahweh.'' And
Solomon prayed unto Yahweh, ''He who dwells in the fog-
like cloud,'' beseeching the Lord ''who dwells in the heavens'' to come and listen to his people's prayers in the new
temple.

The long postponement in the inauguration of the temple
was required, it appears, so that it would take place ''in the
seventh month, on the festival.'' There can be no doubt that
the festival referred to was the New Year festival, in accordance with the commandments concerning holy days and
festivals pronounced in the biblical Book of Leviticus.
''These are the appointed festivals of Yahweh,'' the pream-
ble to chapter 23 states: the observance of the seventh day
as the Sabbath is just the first of holy days to be held in
intervals of multiples of seven days or that were to last
seven days, culminating with the festivals of the seventh
month: New Year's Day, the Day of Atonement, and the
Feast of Booths.

In Mesopotamia by that time Babylon and Assyria had
supplanted Sumer, and the New Year festival was celebrated—as the month's name indicated—in the first month,
called Nissan, which coincided with the spring equinox.
The reasons why the Israelites were commanded to celebrate
the New Year in the seventh month, coinciding with the
autumnal equinox, remain unexplained in the Bible. But we
may find a clue in the fact that the biblical narrative does
not call this month by its Babylonian-Assyrian name, Tishrei, but by the enigmatic name *Etanim*. No satisfactory
explanation for this name has been found so far; but a
solution does occur to us: in view of all the above listed
similarities between the king-priest as a shepherd and the
circumstances of the establishment of a new capital and the
construction of a residence for Yahweh in the desert and in

Jerusalem, the clue to the month's name should also be sought in the *Tale of Etana*. For does not the name used in the Bible, *Etanim*, simply stem from the name *Etana?* The name *Etan* as a personal name, one may note, was not uncommon among the Hebrews, meaning "heroic, mighty."

The celestial alignments in Kish, we have noted, were expressed not only in the temple's solar orientation but also in some relationship with seven planetary "gods" in the heavens. It is noteworthy that in a discussion by August Wünsche of the similarities between Solomon's edifices in Jerusalem and the Mesopotamian "portrait of the heavens" (*Ex Oriente Lux*, vol. 2) he cited the rabbinic reference— as in the *Tale of Etana*—to the "seven stars that indicate time"—Mercury, Moon, Saturn, Jupiter, Mars, Sun, and Venus. There are thus plenty of clues and indications confirming the celestial-calendrical aspects of Solomon's Temple—aspects that link it to traditions and orientations established millennia earlier, in Sumer.

This is reflected not only in the orientation, but also in the temple's tripartite division; it emulated the traditional temple plans that began in Mesopotamia millennia earlier. Günter Martiny, who in the 1930s led the studies regarding the architecture and astronomical orientation of Mesopotamian temples (*Die Gegensätze im Babylonischen und Assyrischen Tempelbau* and other studies) sketched thus (Fig. 45a) the basic tripartite layout of "cult structures": a rectangular anteroom, an elongated ritual hall, and a square Holy of Holies. Walter Andrae (*Des Gotteshaus und die Urformen des Bauens*) pointed out that in Assyria the temple's entrance was flanked by two pylons (Fig. 45b); this was reflected in Solomon's Temple, where the entrance was flanked by two freestanding pillars (see Fig. 44b).

The detailed architectural and construction information in the Bible in respect to Solomon's Temple calls its anteroom *Ulam*, its ritual hall *Hekhal*, and its holiest part *Dvir*. The latter, meaning "Where the speaking takes place," no doubt reflected the fact that Yahweh spoke to Moses from the Ark of the Covenant, the voice coming from where the

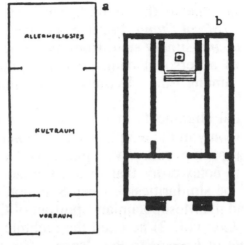

Figures 45a and 45b

wings of the Cherubim were touching; and the Ark was placed in the Temple as the only artifact in the innermost enclosure, the Holy of Holies or *Dvir*. The terminology used for the two foreparts, scholars have recognized, comes from the Sumerian (via Akkadian): *E-gal* and *Ulammu*.

This essential tripartite division, adopted later on elsewhere (e.g. the Zeus temple in Olympia, Fig. 46a, or the Canaanite one at Tainat in Upper Syria, Fig. 46b), was in reality a continuation that began with the most ancient temples, the ziggurats of Sumer, where the way to the ziggurat's top, via a stairway, led through two shrines, an outer shrine with two pylons in front of it, and a prayer room—as drawn by G. Martiny in his studies (Fig. 47).

As in the Sinai Tabernacle and Jerusalem Temple, so were the Mesopotamian vessels and utensils used in the temple rituals made primarily of gold. Texts describing temple rituals in Uruk mention golden libation vessels, golden trays, and golden censers; such objects were found in archaeological excavations. Silver was also used, an example being the engraved vase (Fig. 48) that Entemena, one of the early Sumerian kings, presented to his god Ninurta at the temple in Lagash. The artful votive utensils usually bore

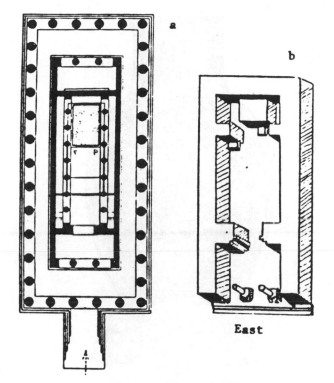

East

Figures 46a and 46b

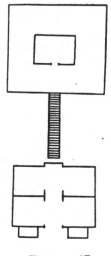

Figure 47

Figure 48

a dedicatory inscription in which the king stated that the object was offered so that the king might be granted long life.

Such presentations could be made only with the permission of the gods, and in many instances were events of great significance, worthy of commemoration in the Date Formulas—listings of the kings' reigns in which each year was named after its main event: the king's ascent to the throne, a war, the presentation of a new temple artifact. Thus, a king of Isin (Ishbi-Erra) called the nineteenth year of his reign "Year in which the throne in the Great House of the goddess Ninlil was made"; and another ruler of Isin (Ishme-Dagan) named one of his regnal years "Year in which Ishme-Dagan made a bed of gold and silver for the goddess Ninlil."

But having been built of bricks made of clay, the temples

of Mesopotamia fell into disrepair as time went by, frequently as the result of earthquakes. Constant maintenance and repairs were required, and repairs or reconstruction of the gods' houses, rather than the offering of new furnishings, began to fill the Date Formulas. Thus, the years-list for the famed Hammurabi, king of Babylon, began with the designation of Year One as the "Year in which Hammurabi became king," and "Year in which the laws were promulgated" for Year Two. Year Four, however, was already designated "Year in which Hammurabi built a wall for the sacred precinct." A successor of Hammurabi in Babylon, the king Shamshi-Iluna, named his eighteenth year as the "Year in which the reconstruction work was done on the E.BABBAR of the god Utu in Sippar" (E.BABBAR, meaning "House of the Bright One," was a temple dedicated to the "Sun-god" Utu/Shamash).

Sumerian, then Akkadian, Babylonian, and Assyrian kings recorded in their inscriptions with great pride how they repaired, embellished, or rebuilt the sacred temples and their precincts; archaeological excavations not only uncovered such inscriptions but also corroborated the claims made therein. In Nippur, for example, archaeologists from the University of Pennsylvania found in the 1880s evidence of repair and maintenance work in the sacred precinct in thirty-five feet of debris piled up during some four thousand years *above* a brick pavement built by the Akkadian king Naram-Sin circa 2250 B.C. and another accumulation of debris of over thirty feet *below* the pavement from earlier times down to virgin soil (which were not excavated and examined at the time).

Returning to Nippur half a century later, a joint expedition of the University of Pennsylvania and the Oriental Institute of the University of Chicago spent many digging seasons working to unearth the Temple of Enlil in Nippur's sacred precinct. The excavators found five successive constructions between 2200 B.C. and 600 B.C., the latter having its floor some twenty feet above the former. The even earlier temples, the archaeologists' report noted at the time, were still

to be dug for. The report also noted that the five temples
were "built one above the other on exactly the same plan."

The discovery that later temples were erected upon the
foundations of earlier temples in strict adherence to the
original plans was reconfirmed at other ancient sites in
Mesopotamia. The rule applied even to enlargement of
temples—even if more than once, as was found at Eridu
(Fig. 49); in all instances the original axis and orientation
were retained. Unlike the Egyptian temples whose solstitial
orientation had to be realigned from time to time because
of the change in the Earth's tilt, Mesopotamian equinoctial
temples needed no adjustment in their orientation because
geographic north and geographic east, by definition, remained unchanged no matter how the Earth's tilt had
changed: the Sun always passed over the equator at "equinox" times, rising on such days precisely in the east.

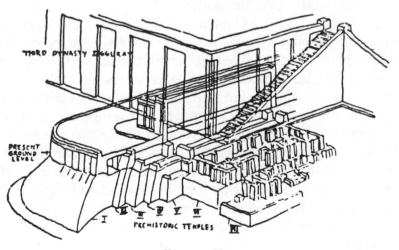

Figure 49

The obligation to adhere to the "olden plans" was spelled
out in an inscription on a tablet found in Nineveh, the
Assyrian capital, among the ruins of a rebuilt temple. In it
the Assyrian king recorded his compliance with the sacred
requirement:

> *The everlasting ground plan,*
> *that which for the future*

the construction determined,
[I have followed.]
It is the one which bears
the drawings from the Olden Times
and the writing of the Upper Heaven.

The Assyrian king Ashur-Nasir-Pal described what such work entailed in a long inscription regarding the restoration of the temple in Calah (an early city mentioned in the Bible). Describing how he had unearthed the "ancient mound," he stated: "I dug down to the level of the water, for 120 measures into the depth I penetrated. I found the foundations of the god Ninib, my lord . . . I constructed thereon, with firm brickwork, the temple of Ninib, my lord." It was done, the king prayed, so that the god Ninib (an epithet for the god Ninurta) "may command that my days be long." Such a blessing, the king hoped, would follow the decision by the god, at a time of his own choosing—"at his heart's desire"—to come and reside in the rebuilt temple: "When the lord Ninib shall take up habitation, forever, in his pure temple, his dwelling place." It is a prayed-for expectation-cum-invitation not unlike the one expressed by King Solomon when the First Temple was completed.

Indeed, the obligatory adherence to the earlier site, orientation, and layout of the temples in the ancient Near East, no matter how long the interval or how extensive the repairs or rebuilding had to be, is exemplified by the successive temples in Jerusalem. The First Temple was destroyed by the Babylonian king Nebuchadnezzar in 587 B.C.; but after Babylon fell to the Achaemenid Persians, the Persian king Cyrus issued an edict permitting the return of Jewish exiles to Jerusalem and the rebuilding of the temple by them. The rebuilding, significantly, began with the erection of an altar (where the first one used to be) "when the seventh month commenced," i.e. on the day of the New Year (and the sacrifices continued until the Feast of Booths). Lest there be doubt about the date, the Book of Ezra (3:6) restated the date: "From the first day of the seventh month did the sacrifices to Yahweh commence."

The adherence not only to the location and orientation of the temple but also to the time of the New Year—an indication of the calendrical aspect of the temple—is re-affirmed in the prophecies of Ezekiel. One of the Jews exiled to Babylon by Nebuchadnezzar, he was shown in a vision the temple-to-come in the New Jerusalem. It happened, the prophet stated (Ezekiel chapter 40) in the month of the New Year, on the 10th of it—precisely the Day of Atonement—that "the hand of Yahweh was upon me, and He brought me thither" (to "the Land of Israel"). "And he sat me up upon a very high mountain, by which was a model of a city." There he saw "a man, his appearance was like the appearance of brass; and he held in his hand a cord of flax and a measuring reed, and he stood in the gate." And this Man of Brass then proceeded to describe the New Temple to Ezekiel. Scholars, using the data, have been able to draw the visionary temple (Fig. 50); it follows precisely the layout and orientation of the temple built by Solomon.

The prophetic vision became a reality after the Persian king Cyrus, having defeated and captured Babylon, issued

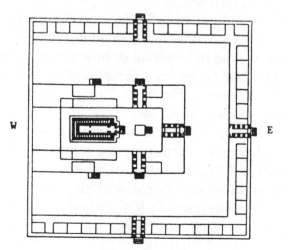

Figure 50

an edict proclaiming the restoration of the destroyed temples throughout the Babylonian empire; a copy of the edict, inscribed on a clay cylinder, has actually been found by archaeologists (Fig. 51). A special royal proclamation, recorded word for word in the Book of Ezra, called on the Jewish exiles to rebuild the "House of Yahweh, God of Heaven."

The Second Temple, built under difficult conditions in what was still a devastated land, was a poor imitation of the First Temple. Rebuilt a part at a time, it was constructed according to plans received from records kept in the Persian royal archives and, the Bible asserts, in strict conformity with the details in the Five Books of Moses. That the Temple indeed followed the original layout and orientation became clearer some five centuries later, when King Herod decided to replace the poor replica with a new, splendid edifice that

Figure 51

would not just match, but even surpass, in grandeur the First Temple. Built on an enlarged great platform (still known as the Temple Mount) and its massive walls (of which the Western Wall, still largely intact, is revered by Jews as the extant remnant of the Holy Temple), it was surrounded by courtyards and various auxiliary buildings. But the House of the Lord proper retained the tripartite layout and orientation of the First Temple (Fig. 52). The Holy of Holies, moreover, remained identical in size to that of the First Temple—and was located *precisely* over its spot; except that the enclosure was no longer called *Dvir,* for the Ark of the Covenant disappeared when the Babylonians destroyed the First Temple and carried off all the artifacts within.

As one views the remains of the immense sacred precincts with their temples and shrines and service buildings, court-yards and gates, and, in the innermost section, the ziggurat, it should be borne in mind that the very first temples were the actual abodes of the gods and were literally called the god's "E"—the god's actual "House." Begun as structures atop artificial mounds and raised platforms (see Fig. 35), they in time evolved to become the famed ziggurats (step-pyramids)—the skyscrapers of antiquity. As an artist's

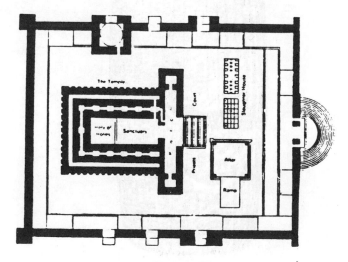

Figure 52

drawing shows (Fig. 53), the deity's actual residence was in the topmost stage. There, seated on their thrones under a canopy, the gods would grant audiences to their chosen king, the "Shepherd of Men." As is shown in this depiction of Utu/Shamash in his temple, the Ebabbar in Sippar (Fig. 54), the king had to be led in by the high priest and was accompanied by his patron god or goddess. (Later on, the

Figure 53

Figure 54

High Priest alone entered the Holy of Holies, as depicted in Fig. 55).

Circa 2300 B.C. a high priestess, the daughter of Sargon of Akkad, collected all the hymns to the ziggurat-temples of her time. Called by Sumerologists "a unique Sumerian literary composition" (A. Sjöberg and E. Bergmann in *Texts From Cuneiform Sources*, vol. 3), the text pays homage to forty-two "E" temples, from Eridu in the south to Sippar in the north and on both sides of the Euphrates and Tigris rivers. The verses not only name the temple, its location, and the god for whom it was built, but also throw light on the magnificence and greatness of these divine abodes as well as on their functions and, sometimes, their history.

The composition appropriately begins with Enki's ziggurat-temple in Eridu, called in the hymn "place whose

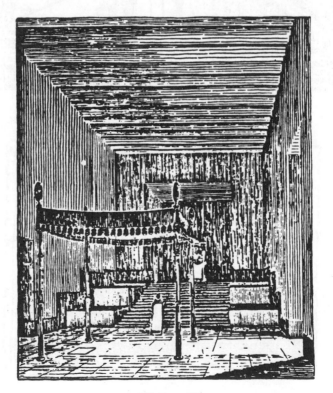

Figure 55

Holy of Holies is the foundation of Heaven-Earth," for Eridu was the first City of the Gods, the first outpost of the first landing party of the Anunnaki (led by Enki), and the first divine city opened up to Earthlings to become also a City of Men. Called E.DUKU, "House of the Holy Mound," it was described in the hymn as a "lofty shrine, rising toward the sky."

This hymn was followed by one to the E.KUR—"House which is like a mountain"—the ziggurat of Enlil in Nippur. Considered the Navel of the Earth, Nippur was equidistant from all the other earliest Cities of the Gods, and was still deemed to be the place from whose ziggurat as one looked to his right he could see Sumer in the south and to his left Akkad in the north, according to the hymn. It was a "shrine where destinies are determined," a ziggurat "which bonds heaven and earth." In Nippur Ninlil, Enlil's spouse, had her separate temple, "clad in awesome brilliance." From it the goddess appeared "in the month of the New Year, on the day of the festival, wonderfully adorned."

The half sister of Enki and Enlil, Ninharsag, who was among the first Anunnaki to come to Earth and was their chief biologist and medical officer, had her temple at the city called Kesh. Simply called E.NINHARSAG. "House of the Lady of the Mountainpeak," it was described as a ziggurat whose "bricks are well moulded . . . a place of Heaven and Earth, an awe inspiring place" which apparently was adorned with "a great poisonous serpent" made of lapis lazuli—the symbol of medicine and healing. (Moses, it will be recalled, made an image of a serpent to stop a killing plague in the Sinai desert).

The god Ninurta, Enlil's Foremost Son by his half sister Ninharsag, who had a ziggurat in his own "cult center," Lagash, had at the time of the composition of this text also a temple in the sacred precinct of Nippur; it was called E.ME.UR.ANNA, "House of the ME's of Anu's Hero." In Lagash, the ziggurat was called E.NINNU, "House of Fifty," reflecting Ninurta's numerical rank in the divine hierarchy (Anu's rank, sixty, was the highest).

It was, the hymn stated, a "House filled with radiance and awe, grown high like a mountain," in which Ninurta's "Black Bird," his flying machine, and his *Sharur* weapon ("the raging storm which envelops men") were housed.

Enlil's firstborn son by his official spouse, Ninlil, was Nannar (later known as Sin), who was associated with the Moon as his celestial counterpart. His ziggurat, in Ur, was called E.KISH.NU.GAL, a "House of Thirty, the great seed" and was described as a temple "whose beaming moonlight comes forth in the land"—all references to Nannar/Sin's celestial association with the Moon and the month.

Nannar/Sin's son, Utu/Shamash (his celestial counterpart was the Sun) had his temple in Sippar, the E.BABBAR— "House of the Bright One" or "Bright House." It was described as "House of the prince of heaven, a heavenly star who from the horizon fills the earth from heaven." His twin sister, Inanna/Ishtar, whose celestial counterpart was the planet Venus, had her ziggurat temple in the city Zabalam, where it was called "House full of brightness"; it was described as a "pure mountain," a "shrine whose mouth opens at dawn" and one "through which the firmament is made beautiful at night"—undoubted reference to the double role of Venus as an evening, as well as a morning, "star." Inanna/Ishtar was also worshiped in Erech, where Anu had put at her disposal the ziggurat-temple built for him when he had come to Earth for a visit. The ziggurat was called E.ANNA, simply "House of Anu." The hymn described it as a "ziggurat of seven stages, surveying the seven luminary gods of the night"—a reference to its alignment and astronomical aspects that was echoed, as we have noted earlier, in rabbinic comments regarding the Jerusalem temple.

Thus did the composition go on, portraying the forty-two ziggurats, their glories, and celestial associations. Scholars speak of this composition from more than 4,300 years ago as a "collection of Sumerian temple hymns" and title it "The Cycle of Old Sumerian Poems about the Great Temples." It may however be much more appropriate to follow the Sumerian custom and call the text by its opening words:

E U NIR   House-ziggurat rising high
AN.KI DA   Heaven-Earth joining.

One of those Houses and its sacred precinct, as we shall see, hold a key that can unlock the Stonehenge enigma and the events of that time's New Age.

5

KEEPERS OF THE SECRETS

Between sunset and sunrise there has been the night.

The Bible constantly saw the Creator's awesomeness in the "Host of Heaven"—the myriads of stars and planets, moons and moonlets that twinkle in the Vault of Heaven as night falls. "The heavens bespeak the glory of the Lord and the vault of heaven reveals His handiwork," the Psalmist wrote. The "heavens" thus described were the nightly skies; and the glory they bespoke was conveyed to Mankind by astronomer-priests. It was they who made sense of the countless celestial bodies, recognized stars by their groups, distinguished between the immovable stars and the wandering planets, knew the Sun's and Moon's movements, and kept track of Time—the cycle of sacred days and festivals, the calendar.

The sacred days began at dusk on the previous evening— a custom still retained in the Jewish calendar. A text which outlined the duties of the *Urigallu* priest during the twelve-day New Year festival in Babylon throws light not only on the origin of priestly rituals later on, but also on the close connection between celestial observations and the festival's proceedings. In the discovered text (generally considered to reflect, as the priest's title URI.GALLU itself, Sumerian origins) the beginning, dealing with the determination of the first day of the New Year (the first of the month Nissan in Babylon) according to the spring equinox, is missing. The inscription starts with the instructions for the second day:

> *On the second day of the month Nisannu,*
> *two hours into the night,*

108

the Urigallu *priest shall arise
and wash with river water.*

Then, putting on a garment of pure white linen, he could
enter into the presence of the great god (Marduk in Babylon)
and recite prescribed prayers in the Holy of Holies of the
ziggurat (the *Esagil* in Babylon). The recitation, which no
one else was to hear, was deemed so secret that after the
text lines in which the prayer was inscribed, the priestly
scribe inserted the following admonition: "Twenty-one
lines: secrets of the Esagil temple. Whoever reveres the god
Marduk shall show them to no one except the Urigallu
priest."

After he finished reciting the secret prayer, the Urigallu
priest opened the temple's gate to let in the *Eribbiti* priests,
who proceeded to "perform their rites, in the traditional
manner," joined by musicians and singers. The text then
details the rest of the duties of the Urigallu priest on that
night.

"On the third day of the month Nisannu" at a time after
sunset too damaged in the inscription to read, the Urigallu
priest was again required to perform certain rites and rec-
itations; this he had to do throughout the night, until "three
hours after sunrise," when he was to instruct artisans in the
making of images of metal and precious stones to be used
in ceremonies on the sixth day. On the fourth day, at "three
and one third hours of the night," the rituals repeated themselves but the prayers now expanded to include a separate
service for Marduk's spouse, the goddess Sarpanit. The
prayers then paid homage to the other gods of Heaven and
Earth and asked for the granting of long life to the king and
prosperity to the people of Babylon. It was thereafter that
the advent of the New Year was directly linked to the Time
of the Equinox in the constellation of the Ram: the heliacal
rising of the Ram Star at dawn. Pronouncing the blessing
"*Iku*-star" upon the "Esagil, image of heaven and earth,"
the rest of the day was spent in prayers, singing and music
playing. On that day, after sunset, the *Enuma elish,* the
Epic of Creation, was recited in its entirety.

The fifth day of Nissan was compared by Henri Frankfort (*Kingship and the Gods*) to the Jewish Day of Atonement, for on that day the king was escorted to the main chapel and was relieved there by the High Priest of all the symbols of kingship; after which, struck in the face by the priest and humiliated into prostrating himself, the king pronounced declarations of confession and repentance. The text which we have been following (per F. Thureau-Dangin, *Rituels accadiens* and E. Ebeling in *Altorientalische Texte zum alten Testament*) deals, however, only with the duties of the Urigallu priest; and we read that on that night the priest, at "four hours of the night," recited twelve times the prayer "My Lord, is he not my Lord" in honor of Marduk, and invoked the Sun, the Moon, and the twelve constellations of the zodiac. A prayer to the goddess followed, in which her epithet, DAM.KI.ANNA ("Mistress of Earth and Heaven") revealed the ritual's Sumerian origin. The prayer likened her to the planet Venus "which shines brilliantly among the stars," naming seven constellations. After these prayers, which stressed the astronomical-calendrical aspects of the occasion, singers and musicians performed "in the traditional manner" and a breakfast was served to Marduk and Sarpanit "two hours after sunrise."

The Babylonian New Year rituals evolved from the Sumerian AKITI ("On Earth Build Life") festival whose roots can be traced to the state visit by Anu and his spouse Antu to Earth circa 3800 B.C., when (as the texts attest) the zodiac was ruled by the Bull of Heaven, the Age of Taurus. We have suggested that it was then that Counted Time, the calendar of Nippur, was granted to Mankind. Inevitably, that entailed celestial observations and thus led to the creation of a class of trained astronomer-priests.

Several texts, some well preserved and some surviving only in fragments, describe the pomp and circumstance of Anu's and Antu's visit to Uruk (the biblical Erech) and the ceremonies which became the rituals of the New Year festival in the ensuing millennia. The works of F. Thureau-Dangin and E. Ebeling still constitute the foundation on which subsequent studies have been based; the ancient texts were then brilliantly used by teams of German excavators of Uruk to locate, identify, and reconstruct the ancient sacred precinct—its walls and gates, its courtyards and shrines and service buildings, and the three principal temples: the E.ANNA ("House of Anu") ziggurat, the *Bit-Resh* ("Main Temple") which was also a stage-tower, and the *Irigal* which was the temple dedicated to Inanna/Ishtar. Of the many volumes of the archaeologists' reports (*Ausgrabungen der Deutschen Forschungsgemeinschaft in Uruk-Warka*), of particular interest to the remarkable correlation of ancient texts and modern excavations are the second (*Archaische Texte aus Uruk*) and third (*Topographie von Uruk*) volumes by Adam Falkenstein.

Surprisingly, the texts on the clay tablets (whose scribal colophons identify them as copies of earlier originals) clearly describe two sets of rituals—one taking place in the month Nissan (the month of the spring equinox) and the other in the month Tishrit (the month of the autumnal equinox); the former was to become the Babylonian and Assyrian New Year, and the latter was retained in the Jewish calendar following the biblical commandment to celebrate the New Year "in the seventh month," *Tishrei*. While the reason for this diversity still mystifies scholars, Ebeling noted that the Nissan texts appear to have been better preserved than the Tishrei texts which are mostly fragmented, suggesting a clear bias on the part of the later temple scribes; and Falkenstein has noted that the Nissan and Tishrei rituals, seemingly identical, were not really so; the former stressed the various celestial observations, the latter the rituals within the Holy of Holies and its anteroom.

Of the various texts, two main ones deal separately with evetime and sunrise rituals. The former, long and well preserved, is especially legible from the point at which Anu and Antu, the divine visitors from Nibiru, are seated in the courtyard of the sacred precinct at evetime, ready to begin a lavish dinner banquet. As the Sun was setting in the west, astronomer-priests stationed on various stages of the main ziggurat were required to watch for the appearance of the planets and to announce the sighting the moment the celestial bodies appeared, beginning with Nibiru:

In the first watch of the night
from the roof of the topmost stage
of the temple-tower of the main temple,
when the planet Great Anu of Heaven,
the planet of Great Antu of Heaven,
shall appear in the constellation Wagon,
the priest shall recite the compositions
Ana tamshil zimu banne kakkab shamami Anu sharru
and Ittatza tzalam banu.

As these compositions ("To the one who grows bright, the heavenly planet of the Lord Anu" and "The Creator's image has arisen") were recited from the ziggurat, wine was served to the gods from a golden libation vessel. Then, in succession, the priests announced the appearance of Jupiter, Venus, Mercury, Saturn, Mars, and the Moon. The ceremony of washing the hands followed, with water poured from seven golden pitchers honoring the six luminaries of the night plus the Sun of daytime. A large torch of "naphtha fire in which spices were inserted" was lighted; all the priests sang the hymn *Kakkab Anu etellu shamame* ("The planet of Anu rose in the sky"), and the banquet could begin. Afterward Anu and Antu retired for the night and leading gods were assigned as watchmen until dawn. Then, "forty minutes after sunrise," Anu and Antu were awakened "bringing to an end their overnight stay."

The morning proceedings began outside the temple, in the courtyard of the *Bit Akitu* ("House of the New Year Festival" in Akkadian). Enlil and Enki were awaiting Anu at the "golden supporter," standing by or holding several objects; the Akkadian terms, whose precise meaning remains elusive, are best translated as "that which opens up the secrets," "the Sun disks" (plural!) and "the splendid/shining posts." Anu then came into the courtyard accompanied by gods in procession. "He stepped up to the Great Throne in the Akitu courtyard, and sat upon it facing the rising Sun." He was then joined by Enlil, who sat on Anu's right, and Enki, who sat on his left; Antu, Nannar/Sin, and Inanna/Ishtar then took places behind the seated Anu.

The statement that Anu seated himself "facing the rising Sun" leaves no doubt that the ceremony involved a determination of a moment connected with sunrise on a particular day—the first day of Nissan (the spring Equinox Day) or the first day of Tishrei (the autumnal Equinox Day). It was only when this sunrise ceremony was completed, that Anu was led by one of the gods and by the High Priest to the BARAG.GAL—the "Holy of Holies" inside the temple.

(BARAG means "inner sanctum, screened-off place" and GAL means "great, foremost." The term evolved to *Baragu/Barakhu/Parakhu* in Akkadian with the meanings "inner sanctum, Holy of Holies" as well as the screen which hides it. This term appears in the Bible as the Hebrew word *Parokhet*, which was both the word for the Holy of Holies in the temple and for the screen that separated it from the anteroom. The traditions and rituals that began in Sumer were thus carried on both physically and linguistically.)

Another text from Uruk, instructing the priests regarding daily sacrifices, calls for the sacrifice of "fat clean rams, whose horns and hooves are whole," to the deities Anu and Antu, "to the planets Jupiter, Venus, Mercury, Saturn and Mars; to the Sun as it rises, and to the Moon on its appearance." The text then explains what "appearance" means in respect to each one of these seven celestial bodies: it meant the moment when they come to rest in the instrument which is "in the midst of the *Bit Mahazzat*" ("House of Viewing"). Further instructions suggest that this enclosure was "on the topmost stage of the temple-tower of the god Anu."

Depictions have been found that show divine beings flanking a temple entrance and holding up poles to which ringlike objects are attached. The celestial nature of the scene is indicated by the inclusion of the symbols of the Sun and the Moon (Fig. 56). In one instance the ancient artist may have intended to illustrate the scene described in the Uruk

Figure 56

Figure 57

ritual text—depicting Enlil and Enki flanking a gateway
through which Anu is making a grand entrance. The two
gods are holding posts to which viewing devices (circular
instruments with a hole in the center) are attached (which
is in accord with the text that spoke of Sun disks in the
plural); the Sun and Moon symbols are shown above the
gateway (Fig. 57).

Other depictions of poles-with-rings freestanding, not
held up, flanking temple entrances (Fig. 58) suggest that

Figure 58

they were the forerunners of the uprights that flanked temples throughout the ancient Near East in ensuing millennia, be it the two columns at Solomon's temple or the Egyptian obelisks. That these originally had an actual and not just symbolic astronomical function could be gathered from an inscription by the Assyrian king Tiglatpileser I (1115–1077 B.C.) in which he recorded the restoration of a temple to Anu and Adad that was built 641 years earlier and that had been lying in ruins for the past sixty years. Describing how he cleared the debris to reach the foundation and followed the original layout in the reconstruction, the Assyrian king said:

> *Two great towers*
> *to discern the two great gods*
> *I built in the House of Brilliance—*
> *a place for their joy,*
> *a place for their pride—*
> *a brilliance of the stars of the heaven.*
> *With the master-builder's artfulness,*
> *with my own planning and exertions,*
> *the insides of the temple I made splendid.*
> *In its midst I made a place for the*

rays directly from the heavens,
in the walls I made the stars to appear.
I made their brilliance great,
the towers I made to rise to the sky.

According to this account, the two great towers of the temple were not just architectural features, but served an astronomical purpose. Walter Andrae, who led some of the most fruitful excavations in Assyria, expressed the view that the serrated ''crowns'' that topped towers that flanked temple gateways in Ashur, the Assyrian capital, indeed served such a purpose (*Die Jüngeren Ishtar-Tempel*). He found confirmation for that conclusion in relevant illustrations on Assyrian cylinder seals, such as in Fig. 59a and 59b, that associate the towers with celestial symbols. Andrae surmised that some of the depicted altars (usually shown to-

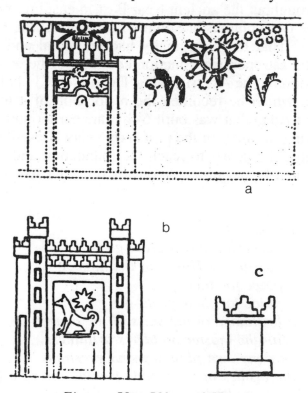

Figures 59a, 59b, and 59c

gether with a priest performing rites) also served a celestial (i.e., astronomical) purpose. In their serrated superstructures (Fig. 59c) these facilities, high up temple gateways or in the open courtyards of temple precincts, created substitutes for the rising stages of the ziggurats as ziggurats gave way to the more easily built flat-roofed temples.

The Assyrian inscription also serves as a reminder that not only the Sun at dawn, and the accompanying heliacal rising of stars and planets, but also the nightly Host of Heaven were observed by the astronomer-priests. A perfect example of such dual observations concerns the planet Venus, which because of its much shorter orbit time around the Sun than Earth's appears to an observer from Earth half the time as an evening star and half the time as a morning star. A Sumerian hymn to Inanna/Ishtar, whose celestial counterpart was the planet we call Venus, offered adoration to the planet first as an evening star, then as a morning star:

> *The holy one stands out in the clear sky;*
> *Upon all the lands and all the people*
> *the goddess looks sweetly from heaven's midst . . .*
>
> *At evetime a radiant star,*
> *a great light that fills the sky;*
> *The Lady of the Evening, Inanna,*
> *is lofty on the horizon.*

After describing how both people and beasts retire for the night "to their sleeping places" after the appearance of the Evening Star, the hymn continues to offer adoration to Inanna/Venus as the Morning Star: "She made the morning come forth, the bright daylight; and in the bedchambers sweet sleep has come to an end."

While such texts throw light on the role of the ziggurats and their rising stages in the observation of the night sky, they also raise the intriguing question: did the astronomer-priests observe the heavens with the naked eye, or did they have instruments for pinpointing the celestial moments of appearances? The answer is provided by depictions of zig-gurats on whose upper stages poles topped by circular ob-jects are emplaced; their celestial function is indicated by the image of Venus (Fig. 60a) or of the Moon (Fig. 60b).

The hornlike devices seen in Fig. 60b serve as a link to Egyptian depictions of instruments for astronomical obser-vations associated with temples. There, viewing devices consisting of a circular part emplaced in the center of a pair of horns atop a high pole (Fig. 61a) were depicted as raised in front of the temples to a god called Min. His festival, celebrated once a year at the time of the summer solstice, involved the erection of a high mast by groups of men pulling cords—a predecessor, perhaps, of the Maypole fes-tival in Europe. Atop the mast are raised the emblems of Min—the temple with the viewing lunar horns (Fig. 61b).

The identity of Min is somewhat of a mystery. The evi-

Figures 60a and 60b

dence suggests that he was already worshiped in predynastic times, even in the archaic period that preceded pharaonic rule by many centuries. Like the earliest Egyptian *Neteru* (''Guardians'') gods, he had come to Egypt from somewhere else. G.A. Wainwright (''Some Celestial Associations of Min'' in the *Journal of Egyptian Archaeology*, vol. XXI) and others believe that he had come from Asia; another opinion (e.g., Martin Isler in the *Journal of the American Research Center in Egypt*, vol. XXVII) was that Min had arrived in Egypt by sea. Min was also known as Amsu or Khem, which according to E.A. Wallis Budge (*The Gods of the Egyptians*) represented the Moon and meant ''regeneration''—a calendrical connotation.

In some Egyptian depictions the Goddess of the Moon, Qetesh, was shown standing next to Min. Even more instructive is Min's symbol (Fig. 61c) that some call his ''dou-

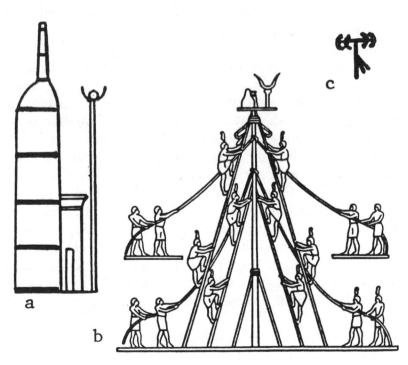

Figures 61a, 61b, and 61c

ble axe weapon'' but others consider to have been a gnomon. We believe that it was a hand-held viewing instrument that represented the crescents of the Moon.

Was Min perhaps another incarnation of Thoth, who was firmly linked to the lunar calendar of Egypt? What is certain is that Min was deemed to be related celestially to the Bull of Heaven, the zodiacal constellation of Taurus, whose age lasted from about 4400 B.C. to about 2100 B.C. The viewing devices that we have seen in the Mesopotamian depictions and those associated with Min in Egypt thus represent some of the oldest astronomical instruments on Earth.

According to the Uruk ritual texts, an instrument called *Itz Pashshuri* was used for the planetary observations. Thureau-Dangin translated the term simply as ''an apparatus''; but the term literally meant an instrument ''that solves, that unlocks secrets.'' Was this instrument one and the same as the circular objects that topped poles or posts, or was the term a generic one, meaning ''astronomical instrument'' in general? We cannot be sure because both texts and depictions have been found, from Sumerian times on, that attest to the existence of a variety of such instruments.

The simplest astronomical device was the *gnomon* (from the Greek ''that which knows''), an instrument which tracked the Sun's movements by the shadow cast by an upright pole; the shadow's length (growing smaller as the Sun rose to midday) indicated the hourly time and the direction (where the Sun's rays first appeared and last cast a shadow) could indicate the seasons. Archaeologists found such devices at Egyptian sites (Fig. 62a) which were premarked to show time (Fig. 62b). Since at solstice times the shadows grew inconveniently long, the flat devices were improved by inclining the horizontal scale, thereby reducing the shadow's length (Fig. 62c). In time, this led to actual structural shadow clocks which were built as stairways that indicated time as the shadow moved up or down the stairs (Fig. 62d).

Shadow clocks also developed into sundials when the upright pole was provided with a semicircular base on which an angular scale was marked. Archaeologists have found

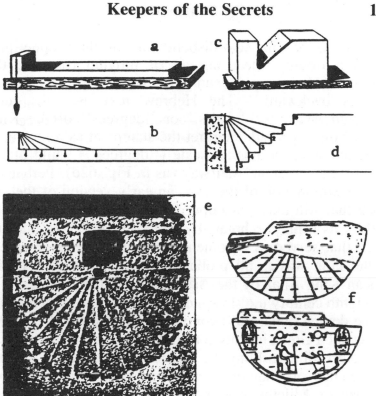

Figures 62a, 62b, 62c, 62d, 62e, and 62f

such instruments at Egyptian sites (Fig. 62e), but the oldest device so far discovered comes from the Canaanite city of Gezer in Israel; it has the regular angular scale on its face and a scene of the worship of the Egyptian god Thoth on the reverse side (Fig. 62f). This sundial, made of ivory, bears the cartouche of the Pharaoh Merenptah, who reigned in the thirteenth century B.C.

Shadow clocks are mentioned in the Bible. The Book of Job refers to portable gnomons, probably of the kind shown in 62a, that were used in the fields to tell time, when it observes that the hired laborer "earnestly desireth the shadow" that indicated it was time to collect his daily wages (Job 7:2). Less clear is the nature of a shadow clock that featured in a miraculous incident reported in II Kings chapter 20 and Isaiah chapter 38. When the prophet Isaiah told the ailing King Hezekiah that he would fully recover within three days, the king was disbelieving. So the prophet predicted a divine omen: instead of moving forward, the shadow of the Temple's sun clock would be "brought ten degrees backward." The Hebrew text uses the term *Ma'aloth Ahaz,* the "stairs" or "degrees" of the King Ahaz. Some scholars interpret the statement as referring to a sundial with an angular scale ("degrees") while others think it was an actual stairway (as in Fig. 62d). Perhaps it was a combination of the two, an early version of the sun clock that still exists in Jaipur, India (Fig. 63).

Be it as it may, scholars by and large agree that the sun clock that served as an omen for the miraculous recovery of the king was in all probability a gift presented to the Judean King Ahaz by the Assyrian king Tiglatpileser II in the eighth century B.C. In spite of the Greek name (*gnomon*) of the device (whose use continued into the Middle Ages), it was not a Greek invention nor, it seems, even an Egyptian one. According to Pliny the Elder, the first-century savant, the science of gnomonics was first described by Anaximander of Miletus who possessed an instrument called "shadow hunter." But Anaximander himself, in his work (in Greek) *Upon Nature* (547 B.C.) wrote that he had obtained the gnomon from Babylon.

The text in II Kings chapter 20, it seems to us, suggests a sundial rather than a built staircase and that it was placed in the Temple courtyard (it had to be in the open where the

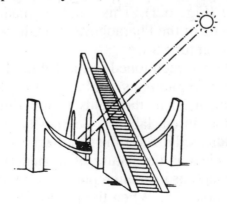

Figure 63

Sun could cast shadows). If Andrae was right regarding the astronomical function of altars, it was possible that the instrument was placed upon the Temple's main altar. Such altars had four "horns," a Hebrew term (*Keren*) that also meant "corner" as well as "beam, ray"—terms suggesting a common astronomical origin. Pictorial evidence supporting such a possibility ranges from early depictions of ziggurats in Sumer, where "horns" preceded the circular objects (Fig. 64a) all the way to Greek times. In tablets depicting altars from several centuries after Hezekiah's time, we can see (Fig. 64b) a viewing ring on a short support placed between two altars; in a second illustration (Fig. 64c) we can see an altar flanked by devices for Sun viewing and Moon viewing.

In considering the astronomical instruments of antiquity, we are in fact dealing with knowledge and sophistication

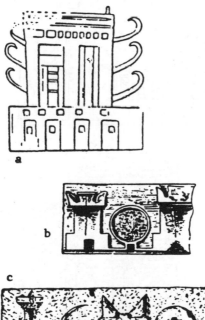

Figures 64a, 64b, and 64c

that go back millennia to ancient Sumer. One of the most archaic depictions from Sumer that shows a procession of temple attendants holding tools and instruments, pictures one of them holding a pole surmounted by an astronomical instrument: a device that connects two short posts with sighting rings on their tops (Fig. 65a). The twin rings in such an arrangement are familiar even nowadays in modern binoculars or theodolites for creating and measuring depth and distance. By carrying it, the attendant makes clear that it was a portable device, an instrument that could be emplaced in various viewing positions.

If the process of celestial observation progressed from massive ziggurats and great stone circles to lookout towers and specially designed altars, the instruments with which the astronomer-priests scanned the heavens at night or tracked the Sun in daytime must have progressed in tandem.

a

b

Figures 65a and 65b

That such instruments became portable thus makes much sense, especially if some were used not only for the original calendrical purposes (fixing festival times) but also for navigation. By the end of the second millennium B.C. the Phoenicians of northern Canaan had become the best navigators of the ancient world; plying the trade routes, one may say, between the stone pillars of Byblos and the ones in the British Isles, their foremost western outpost was Carthage (*Keret-Hadash*, "New City"). There they adopted as their main divine symbol the depiction of an astronomical instrument; before it began to appear on stelae and even tombstones, it was shown in association with two double-ringed pillars that flanked the entrance to a temple (Fig. 65b)—as earlier in Mesopotamia. The ring flanked by two opposite-facing crescents suggests observations of the Sun and of the Moon's phases.

A "votive tablet" found in the ruins of a Phoenician settlement in Sicily (Fig. 66a) depicts a scene in an open courtyard, suggesting that the Sun's movements rather than the night sky are the astronomical objective. The ringed pillar and an altar stand in front of a three-columned structure; there, too, is the viewing device: a ring between two short vertical posts placed on a horizontal bar and mounted atop a triangular base. This particular shape for observations of the Sun brings to mind the Egyptian hieroglyph for "horizon"—the Sun rising between two mountains (Fig. 66b). Indeed, the Phoenician device (scholars refer to it as a "cult symbol") suggesting a pair of raised hands is related to the Egyptian hieroglyph for *Ka* (Fig. 66c) that represented the pharaoh's spirit or alter ego for the Afterlife journey to the abode of the gods on the "Planet of millions of years." That the origin of the *Ka* was, to begin with, an astronomical instrument is suggested by an archaic Egyptian depiction (Fig. 66d) of a viewing device in front of a temple.

All these similarities and their astronomical origin should add new insights to understanding Egyptian depictions (Fig. 67) of the Ka's ascent toward the gods' planet with outstretched hands that emulate the Sumerian device; it ascends from atop a pillar equipped with gradation-steps.

Figures 66a, 66b, 66c, and 66d

Figure 67

The Egyptian hieroglyph depicting this step-pillar was called *Ded,* meaning "Everlastingness." It was often shown in pairs, because two such pillars were said to have stood in front of the principal temple to the great Egyptian god Osiris in Abydos. In the Pyramid Texts, in which the pharaohs' afterlife journeys are described, the two *Ded* pillars are shown flanking the "Door of Heaven." The double doors stay closed until the arriving alter ego of the king pronounces the magical formula: "O Lofty One, thou Door of Heaven: the king has come to thee; cause this door to be opened for him." And then, suddenly, the "double doors of heaven open up . . . the aperture of the celestial windows is open." And, soaring as a great falcon, the pharaoh's *Ka* has joined the gods in Everlastingness.

The Egyptian Book of the Dead has not reached us in the form of a cohesive book, assuming that a composition that might be called a "book" truly existed; rather, it has been collated from the many quotations from it that cover the walls of royal tombs. But a complete book did reach us from ancient Egypt, and it shows that an ascent heavenward to attain immortality was deemed connected with the calendar.

The book we refer to is the *Book of Enoch,* an ancient composition known from two sets of versions, an Ethiopic one that scholars identify as "1 Enoch," and a Slavonic version that is identified as "2 Enoch," and which is also known as *The Book of the Secrets of Enoch.* Both versions, of which copied manuscripts have been found mostly in Greek and Latin translations, are based on early sources that enlarged on the short biblical mention that Enoch, the seventh Patriarch after Adam, did not die because, at age 365, "he walked with God"—taken heavenward to join the deity.

Enlarging on this brief statement in the Bible (Genesis chapter 5), the books describe in detail Enoch's two celestial journeys—the first to learn the heavenly secrets, return, and impart the knowledge to his sons; and the second to stay put in the heavenly abode. The various versions indicate wide astronomical knowledge concerning the motions of the

Sun and the Moon, the solstices and the equinoxes, the reasons for the shortening and lengthening days, the structure of the calendar, the solar and lunar years, and the rule of thumb for intercalation. In essence, the secrets that were imparted to Enoch and by him to his sons to keep, were the knowledge of astronomy as it related to the calendar.

The author of *The Book of the Secrets of Enoch,* the so-called Slavonic version, is believed to have been (to quote R.H. Charles, *The Apocrypha and Pseudepigrapha of the Old Testament*) ''a Jew who lived in Egypt, probably in Alexandria'' some time around the beginning of the Christian era. This is how the book concludes:

Enoch was born on the sixth day of the month Tsivan, and lived three hundred and sixty-five years.

He was taken up to heaven on the first day of the month Tsivan and remained in heaven sixty days. He wrote all these signs of all creation which the Lord created, and wrote three hundred and sixty-six books, and handed them over to his sons.

He was taken up [again] to heaven on the sixth day of the month Tsivan, on the very day and hour when he was born.

Methosalam and his brethren, all the sons of Enoch, made haste and erected an altar at the place called Ahuzan, whence and where Enoch had been taken up to heaven.

Not only the contents of the *Book of Enoch*—astronomy as it relates to the calendar—but also the very life and ascension of Enoch are thus replete with calendrical aspects. His years on Earth, 365, are of course the number of whole days in a solar year; his birth and departure from Earth are linked to a specific month, even the day of the month.

The Ethiopic version is deemed by scholars to be older by several centuries than the Slavonic one, and portions of that older version are in turn known to have been based on even older manuscripts, such as a lost *Book of Noah*. Fragments of Enoch books were discovered among the Dead Sea scrolls. The astronomical-calendrical tale of Enoch thus goes back into great antiquity—perhaps, as the Bible asserts, to pre-Diluvial times.

Now that it is certain that the biblical tales of the Deluge and the *Nefilim* (the biblical Anunnaki), of the creation of the Adam and of Earth itself, and of ante-Diluvial patriarchs, are abbreviated renderings of original earlier Sumerian texts that recorded all that, it is almost certain that the biblical "Enoch" was the equivalent of the Sumerian first priest, EN.ME.DUR.AN.KI ("High Priest of the ME's of the Bond Heaven-Earth"), the man from the city Sippar taken heavenward to be taught the secrets of Heaven and Earth, of divination, and of the calendar. It was with him that the generations of astronomer-priests, of Keepers of the Secrets, began.

The granting by Min to the Egyptian astronomer-priests of the viewing device was not an extraordinary action. A Sumerian sculpture molded in relief shows a great god granting a hand-held astronomical device to a king-priest (Fig. 68). Numerous other Sumerian depictions show a king being granted a measuring rod and a rolled measuring cord for the purpose of assuring the correct astronomical orientation of temples, as we have seen in Fig. 54. Such depictions only enhance the textual evidence that is explicit about the manner in which the line of astronomer-priests began.

Did Man, however, become haughty enough to forget all that, to start thinking he had attained all that knowledge by himself? Millennia ago the issue was tackled when Job was asked to admit that not Man but *El*, "the Lofty One," was the Keeper of the Secrets of Heaven and Earth:

> *Say, if thou knowest science:*
> *Who hath measured the Earth, that it be known?*
> *Who hath stretched a cord upon it?*
> *By what were its platforms wrought?*
> *Who hath cast its Stone of Corners?*

Figure 68

Have you ever ordered Morning or figured out Dawn according to the corners of the Earth? Job was asked. Do you know where daylight and darkness exchange places, or how snow and hailstones come about, or rains, or dew? "Do you know the celestial laws, or how they regulate that which is upon the Earth?"

The texts and depictions were intended to make clear that the human Keepers of the Secrets were pupils, not teachers. The records of Sumer leave no doubt that the teachers, the original Keepers of the Secrets, were the Anunnaki.

The leader of the first team of Anunnaki to come to Earth, splashing down in the waters of the Persian Gulf, was E.A— he "whose home is water." He was the chief scientist of the Anunnaki and his initial task was to obtain the gold they needed by extracting it from the gulf's waters—a task requiring knowledge in physics, chemistry, metallurgy. As a shift to mining became necessary and the operation moved to southeastern Africa, his knowledge of geography, geology, geometry—of all that we call Earth Sciences—came into play; no wonder his epithet-name changed to EN.KI, "Lord Earth," for his was the domain of Earth's secrets. Finally, suggesting and carrying out the genetic engineering that brought the Adam into being—a feat in which he was helped by his half sister Ninharsag, the Chief Medical Officer—he demonstrated his prowess in the disciplines of Life Sciences: biology, genetics, evolution. More than one hundred ME's, those enigmatic objects that, like computer disks, held the knowledge arranged by subject, were kept by him in his center, Eridu, in Sumer; at the southern tip of Africa, a scientific station held "the tablet of wisdom."

All that knowledge was in time shared by Enki with his six sons, each of whom became expert in one or more of these scientific secrets.

Enki's half brother EN.LIL—"Lord of the Command"— arrived next on Earth. Under his leadership the number of Anunnaki on Earth increased to six hundred; in addition, three hundred IGI.GI ("Those who observe and see") remained in Earth orbit, manning orbiting stations, operating shuttlecraft to and from spacecraft. He was a great spaceman, organizer, disciplinarian. He established the first Mission Control Center in NI.IBRU, known to us by its Akkadian name Nippur, and the communication links with the Home Planet, the DUR.AN.KI—"Bond Heaven-Earth." The space charts, the celestial data, the secrets of astronomy were his to know and keep. He planned and supervised the setting up of the first space base in Sippar ("Bird City"). Matters of weather, winds and rains, were his concern; so was the assurance of efficient transportation and supplies, including the local provision of foodstuffs and the arts and crafts of agriculture and shepherding. He maintained discipline among the Anunnaki, chaired the council of the "Seven who Judge," and remained the supreme god of law and order when Mankind began to proliferate. He regulated the functions of the priesthood, and when kingship was instituted, it was called by the Sumerians "Enlilship."

A long and well-preserved *Hymn to Enlil, the All Beneficent* found among the ruins of the E.DUB.BA, "House of Scribal Tablets" in Nippur, mentioned in its one hundred seventy lines many of the scientific and organizational achievements of Enlil. On his ziggurat, the E.KUR ("House which is like a mountain"), he had a "beam that searched the heart of all the lands." He "set up the *Duranki*," the "Bond Heaven-Earth." In Nippur, "a bellwether of the universe" he erected. Righteousness and justice he decreed. With "ME's of heaven" that "none can gaze upon" he established in the innermost part of the Ekur "a heavenly zenith, as mysterious as the distant sea," containing the "starry emblems . . . carried to perfection"; these enabled the establishment of rituals and festivals. It was under Enlil's guidance that "cities were built, settlements founded, stalls built, sheepfolds erected," riverbanks controlled for overflowing, canals built, fields and meadows "filled with rich grain," gardens made to produce fruits, weaving and entwining taught.

Those were the aspects of knowledge and civilization that Enlil bequeathed to his children and grandchildren, and through them to Mankind.

The process by which the Anunnaki imparted such diverse aspects of science and knowledge to Mankind has been a neglected field of study. Little has been done to pursue, for example, such a major issue as how astronomer-priests came into being—an event without which we, today, would neither know much about our own Solar System nor be able to venture into space. Of the pivotal event, the teaching of the heavenly secrets to Enmeduranki, we read in a little-known tablet that was fortunately brought to light by W. G. Lambert in his study *Enmeduranki and Related Material* that

> *Enmeduranki [was] a prince in Sippar,*
> *Beloved of Anu, Enlil and Ea.*
> *Shamash in the Bright Temple appointed him [as*
>   *priest].*
> *Shamash and Adad [took him] to the assembly [of the*
>   *gods] . . .*

They showed him how to observe oil on water,
a secret of Anu, Enlil and Ea.
They gave him the Divine Tablet,
the kibbu *secret of Heaven and Earth . . .*
They taught him how to make calculations with
*   numbers.*

When the instruction of Enmeduranki in the secret knowledge of the Anunnaki was accomplished, he was returned to Sumer. The "men of Nippur, Sippar and Babylon were called into his presence." He informed them of his experiences and of the establishment of the institution of priesthood and that the gods commanded that it should be passed from father to son:

The learned savant
who guards the secrets of the gods
will bind his favored son with an oath
before Shamash and Adad . . .
and will instruct him in the secrets of the gods.

The tablet has a postcript:

Thus was the line of priests created,
those who are allowed to approach Shamash and Adad.

According to the Sumerian King Lists Enmeduranna was the seventh pre-Diluvial holder of kingship, and reigned in Sippar during six orbits of Nibiru before he became High Priest and was renamed Enmeduranki. In the *Book of Enoch* it was the archangel Uriel ("God is my light") who showed Enoch the secrets of the Sun (solstices and equinoxes, "six portals" in all) and the "laws of the Moon" (including intercalation), and the twelve constellations of the stars, "all the workings of heaven." And in the end of the schooling, Uriel gave Enoch—as Shamash and Adad had given Enmeduranki—"heavenly tablets," instructing him to study them carefully and note "every individual fact" therein. Returning to Earth, Enoch passed this knowledge to his oldest son, Methuselah. The *Book of the Secrets of Enoch* includes in the knowledge granted Enoch "all the workings of heaven, earth and the seas, and all the elements, their passages and goings and the thunderings of the thunder; and of the Sun and the Moon; the goings and changings of the stars; the seasons, years, days, and hours." This would be in line with the attributes of Shamash—the god whose celestial counterpart was the Sun and who commanded the spaceport, and of Adad who was the "weather god" of antiquity, the god of storms and rains. Shamash (Utu in Sumerian) was usually depicted (see Fig. 54) holding the measuring rod and cord; Adad (Ishkur in Sumerian) was shown holding the forked lightning. A depiction on the royal seal of an Assyrian king (Tukulti-Ninurta I) shows the king being introduced to the two great gods, perhaps for the purpose of granting him the knowledge they had once given to Enmeduranki (Fig. 69).

Appeals by later kings to be granted as much "Wisdom" and scientific knowledge as renowned early sages had possessed, or boasts that they knew as much, were not uncommon. Royal Assyrian correspondence hailed a king as "surpassing in knowledge all the wise men of the Lower World" because he was an offspring of the "sage Adapa." Another instance had a Babylonian king claim that he possessed "wisdom that greatly surpassed even that contained in the writings that Adapa had composed." These were

Figure 69

references to Adapa, the Sage of Eridu (Enki's center in Sumer), whom Enki had taught "wide understanding" of "the designs of Earth"—the secrets of Earth Sciences.

One cannot rule out the possibility that, as Enmeduranki and Enoch, Adapa too was the seventh in a line of sages, the Sages of Eridu, and thus another version of the Sumerian memory echoed in the biblical Enoch record. According to this tale, seven Wise Men were trained in Eridu, Enki's city; their epithets and particular knowledge varied from version to version. Rykle Borger, examining this tale in light of the Enoch traditions (*"Die Beschworungsserie Bit Meshri und die Himmelfahrt Henochs"* in the *Journal of Near Eastern Studies,* vol. 33), was especially fascinated by the inscription on the third tablet of the series of Assyrian Oath Incantations. In it the name of each sage is given and his main call on fame is explained; it says thus of the seventh: "Utu-abzu, he who to heaven ascended." Citing a second such text, R. Borger concluded that this seventh sage, whose name combined that of Utu/Shamash with the Lower World (*Abzu*) domain of Enki, was the Assyrian "Enoch."

According to the Assyrian references to the wisdom of Adapa, he composed a book of sciences titled U.SAR d ANUM d ENLILA—"Writings regarding Time; from divine Anu and divine Enlil." Adapa, thus, is credited with writing Mankind's first book of astronomy and the calendar.

When Enmeduranki ascended to heaven to be taught the various secrets, his patron gods were Utu/Shamash and Adad/Ishkur, a grandson and a son of Enlil. His ascent was thus under Enlilite aegis. Of Adapa we read that when Enki sent him heavenward to Anu's abode, the two gods who acted as his chaperons were Dumuzi and Gizzida, two sons of Ea/Enki. There, "Adapa from the horizon of heaven to the zenith of heaven cast a glance; he saw its awesomeness"—words reflected in the Books of Enoch. At the end of the visit Anu denied him everlasting life; instead, "the priesthood of the city of Ea to glorify in future" he decreed for Adapa.

The implication of these tales is that there were two lines of priesthood—one Enlilite and one Enki'ite; and two central scientific acadamies, one in Enlil's Nippur and the other in Enki's Eridu. Both competing and cooperating, no doubt, as the two half brothers themselves were, they appear to have acquired their specialties. This conclusion, supported by later writings and events, is reflected in the fact that we find the leading Anunnaki having each their talents, specialty, and specific assignments.

As we continue to examine these specialties and assignments, we will find that the close relationship of temple-astronomy-calendar was also expressed in the fact that several deities, in Sumer as in Egypt, combined these specialties in their attributes. And since the ziggurats and temples served as observatories—to determine the passage of both Earthly and Celestial Times—the deities with the astronomical knowledge were also the ones with the knowledge of orienting and designing the temples and their layouts.

"Say, if thou knowest science: Who hath measured the Earth, that it be known? Who hath stretched a *cord* upon it?" So was Job asked when called upon to admit that God, not Man, was the ultimate Keeper of the Secrets. In the scene of the introduction of the king-priest to Shamash (Fig. 54), the purpose or essence of the occurrence is indicated by two Divine Cordholders. The two cords they stretch to a ray-emitting planet form an angle, suggesting measurement not so much of distance as of orientation. An Egyptian depiction of a similar motif, a scene painted on the Papyrus of Queen Nejmet, shows how two cordholders measured an angle based on the planet called "Red Eye of Horus" (Fig. 70).

The stretching of cords to determine the proper astronomical orientation of a temple was the task of a goddess called Sesheta in Egypt. She was on the one hand a Goddess of the Calendar; her epithets were "the great one, lady of letters, mistress of the House of Books" and her symbol was the stylus made of a palm branch, which in Egyptian hieroglyphs stood for "counting the years." She was depicted with a seven-rayed star within the Heavenly Bow on her head. She was the Goddess of Construction, but only (as pointed out by Sir Norman Lockyer in *The Dawn of*

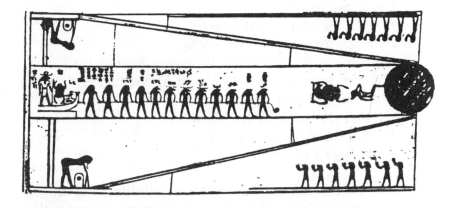

Figure 70

Astronomy) for the purpose of determining the orientation of temples. Such an orientation was not haphazard or a matter left to guesswork. The Egyptians relied on divine guidance to determine the orientation and major axis of their temples; the task was assigned to Sesheta. Auguste Mariette, reporting on his finds at Denderah where depictions and inscriptions pertaining to Sesheta were discovered, said it was she who "made certain that the construction of sacred shrines was carried out exactly according to the directions contained in the Divine Books."

Determining the correct orientation called for an elaborate ceremony named *Put-ser,* meaning "the stretching of the cord." The goddess sank a pole in the ground by hammering it down with a golden club; the king, guided by her, sank another pole. A cord was then stretched between the two poles, indicating the proper orientation; it was determined by the position of a specific star. A study by Z. Zaba published by the Czechoslovak Academy of Sciences (*Archiv Orientalni,* Supplement 2, 1953) concluded that the ceremony revealed knowledge of the phenomenon of precession, and thus of the zodiacal division of the celestial circle. The astral aspects of the ceremony have been made clear by relevant inscriptions, as the one found on the walls of the temple of Horus in Edfu. It records the words of the pharaoh:

I take the peg-pole,
I grasp the club by its handle,
I stretch the cord with Sesheta.
I turn my sight to follow the stars' movements,
I fix my gaze on the astrality of Msihettu.
The star-god that announces the time
reaches the angle of its Merkhet;
I establish the four corners
of the god's temple.

In another instance concerning the rebuilding of a temple in Abydos by the Pharaoh Seti I, the inscription quotes the king thus:

The hammering club in my hand was of gold.
I struck the peg with it.
Thou wast with me in thy capacity of Harpedonapt.
Thy hand held the spade during the fixing of
the temple's four corners with accuracy
by the four supports of heaven.

The ceremony was pictorially depicted on the temple's walls (Fig. 71).

Figure 71

Sesheta was, according to Egyptian theology, the female companion and chief assistant of Thoth, the Egyptian god of sciences, mathematics, and the calendar—the Divine Scribe, who kept the gods' records, and the Keeper of the Secrets of the construction of the pyramids.

As such, he was the foremost Divine Architect.

6

THE DIVINE ARCHITECTS

Some time between 2200 and 2100 B.C.—a time of great import at Stonehenge—Ninurta, Enlil's Foremost Son, embarked on a major undertaking: the building of a new "House" for himself at Lagash.

The event throws light on many matters of gods and men thanks to the fact that the king entrusted with the task, Gudea of Lagash, wrote it all down in great detail on two large clay cylinders. In spite of the immensity of the task, he realized that it was a great honor and a unique opportunity to have his name and deeds remembered for all time, for not many kings were so entrusted; in fact, royal records (since found by archaeologists) spoke of at least one instance when a famous king (Naram-Sin), otherwise beloved by the gods, was again and again refused permission to engage in the building of a new temple (such a situation arose millennia later in the case of King David in Jerusalem). Shrewdly expressing his gratitude to his god by inscribing laudatory statements on statues of himself (Fig. 72) which Gudea then emplaced in the new temple, Gudea managed to leave behind a rather substantial amount of written information which explains the How and What for of the sacred precincts and temples of the Anunnaki.

As the Foremost Son of Enlil by his half sister Ninharsag and thus the heir apparent, Ninurta shared his father's rank of fifty (that of Anu being the highest, sixty, and that of Anu's other son, Enki, being forty) and so it was a simple choice to call Ninurta's ziggurat E.NINNU, the "House of Fifty."

Throughout the millennia Ninurta was a faithful aide to his father, carrying out dutifully each task assigned to him. He acquired the epithet "Foremost Warrior of Enlil" when

Figure 72

a rebel god called Zu seized the Tablets of Destinies from Mission Control Center in Nippur, disrupting the bond Heaven-Earth; it was Ninurta who pursued the usurper to the ends of Earth, seizing him and restoring the crucial tablets to their rightful place. When a brutal war, which in *The Wars of Gods and Men* we have called the Second Pyramid War, broke out between the Enlilites and the Enki'ites, it was again Ninurta who led his father's side to victory. That conflict ended with a peace conference forced on the warring clans by Ninharsag, in the aftermath of which the Earth was divided among the two brothers and their sons and civilization was granted to Mankind in the "Three Regions"—Mesopotamia, Egypt, and the Indus Valley.

The peace that ensued lasted a long time, but not forever. One who had been unhappy with the arrangements all along was Marduk, the Firstborn Son of Enki. Reviving the rivalry between his father and Enlil which stemmed from the complex succession rules of the Anunnaki, Marduk challenged the grant of Sumer and Akkad (what we call Mesopotamia) to the offspring of Enlil and claimed the right to a Mesopotamian city called *Bab-Ili* (Babylon)—literally, "Gateway of the Gods." As a result of the ensuing conflicts, Marduk was sentenced to be buried alive inside the Great Pyramid of Giza; but, pardoned before it was too late, he was forced into exile; and once again Ninurta was called upon to help resolve the conflicts.

Ninurta, however, was not just a warrior. In the aftermath of the Deluge it was he who dammed the mountain passes to prevent more flooding in the plain between the Euphrates and Tigris rivers and who had arranged extensive drainage works there to make the plain habitable again. Thereafter, he oversaw the introduction of organized agriculture to the region and was fondly nicknamed by the Sumerians *Urash*— "He of the Plough." When the Anunnaki decided to give Kingship to Mankind, it was Ninurta who was assigned to organize it at the first City of Men, Kish. And when, after the upheavals caused by Marduk, the lands quieted down circa 2250 B.C., it was again Ninurta who restored order and kingship from his "cult city," Lagash.

His reward was permission from Enlil to build a brand new temple in Lagash. It was not that he was "homeless"; he already had a temple in Kish, and a temple within the sacred precinct in Nippur, next to his father's ziggurat. He also had his own temple in the *Girsu,* the sacred precinct of his "cult center," the city of Lagash. French teams of archaeologists who have been excavating at the site (now locally called Tello), conducting twenty "campaigns" between 1877 and 1933, uncovered many of the ancient remains of a square ziggurat and rectangular temples whose corners were precisely oriented to the cardinal points (Fig. 73). They estimated that the foundations of the earliest temple were laid in Early Dynastic times, before 2700 B.C., on the mound marked "K" on the excavations map. Inscriptions by the earliest rulers of Lagash already spoke of rebuilding and improvements in the Girsu, as well as of the

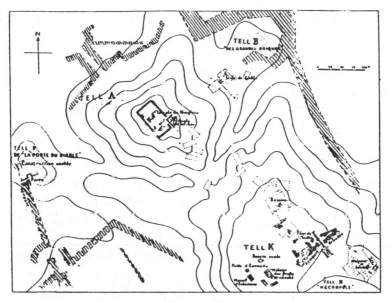

Figure 73

Figure 74

presentation of votive artifacts, such as the silver vase by Entemena (Fig. 48), over a period of six or seven hundred years before Gudea's time. Some inscriptions may mean that the foundations for the very first Eninnu were laid by Mesilim, a king of Kish who had reigned circa 2850 B.C.

Kish, it will be recalled, was where Ninurta had established for the Sumerians the institution of Kingship. For a long time the rulers of Lagash were considered just viceroys, who had to earn the title "king of Kish" in order to be full-fledged sovereigns. It was perhaps this second-class stigma that made Ninurta seek a temple truly authentic for his city; he also needed one that could hold the remarkable weapons that he had been granted by Anu and Enlil, including an aircraft that was nicknamed the Divine Storm Bird (Fig. 74) which had a wingspan of about seventy-five feet and thus needed a specially designed enclosure.

When Ninurta defeated the Enki'ites he entered the Great Pyramid and for the first time realized its intricate and amazing inner architecture in addition to its outer grandeur. The information provided by the inscriptions of Gudea suggests that Ninurta had nourished a desire to have a ziggurat of similar greatness and intricacy ever since his Egyptian tour of duty. Now that he had once again pacified Sumer and attained for Lagash the status of a royal capital, he asked Enlil once again for permission to build a new E.NINNU, a new "House of Fifty," in the Girsu precinct of Lagash. This time, his wish was to be fulfilled.

That his wish was granted should not be downplayed as a matter of course. We read, for example, in the Canaanite "myths" regarding the god Ba'al ("Lord"), that for his role in defeating the enemies of El ("The Lofty One," the supreme deity) he sought El's permission to build a House on the crest of Mount Zaphon in Lebanon. Ba'al had sought this permission before, and was repeatedly turned down; he had repeatedly complained to "Bull El, his father":

> No house has Ba'al like the gods,
> no precinct like the children of Asherah;
> the abode of El is the shelter of his son.

Now he asked Asherah, El's spouse, to intercede for him; and Asherah finally convinced El to give his permission. Added to the previous arguments was a new one: Ba'al, she said, could then "observe the seasons" in his new House— make there celestial observations for a calendar.

But though a god, Ba'al could not just go ahead and build his temple-abode. The plans had to be drawn and construction supervised by the *Kothar-Hasis,* the "Skilled and Knowing" Craftsman of the Gods. Not only modern scholars but even Philo of Byblos in the first century A.D. (quoting earlier Phoenician historians) compared Kothar-Hasis with the Greek divine craftsman Hephaestus (who built the temple-abode of Zeus) or with Thoth, the Egyptian god of knowledge, crafts, and magic. The Canaanite texts indeed state that Ba'al sent emissaries to Egypt to fetch Kothar-Hasis, but found him eventually in Crete.

No sooner, however, did Kothar-Hasis arrive than Ba'al got into fierce arguments with him regarding the temple's architecture. He wanted, it appears, a House of only two parts, not the customary three—an *Hekhal* and a *Bamtim* (a raised stage). The sharpest argument was over a funnellike window or skylight which Kothar-Hasis claimed had to be positioned "in the House" but Ba'al vehemently argued should be located somewhere else. The argument is given the space of many verses in the text to show its ferocity and importance; it involved shouting and spitting . . .

The reasons for the argument regarding the skylight and its location remain obscure; our guess is that it might have been connected with the temple's orientation. The statement by Asherah that the temple would enable observance of the seasons suggests an orientation requiring certain astronomical observations. Ba'al, on the other hand, as the Canaanite text later reveals, was planning to install in the temple a secret communication device that would enable him to seize power over other gods. To that purpose Ba'al "stretched a cord, strong and supple," from the peak of Zaphon ("North") to Kadesh ("the Sacred Place") in the south, in the Sinai desert.

The orientation in the end remained the way the divine architect, Kothar-Hasis, wanted it. "Thou shalt heed my words," he told Ba'al emphatically, and "as for Ba'al, his house was thus built." If, as one must assume, the later temples atop the Baalbek platform were built according to that olden plan, then we find that the orientation Kothar-Hasis had insisted upon resulted in a temple with an east–west axis (see Fig. 25).

As the Sumerian tale of the new Eninnu temple unfolds, we shall see that it too involved celestial observations to determine its orientation, and required the services of divine architects.

* * *

Much like King Solomon some thirteen hundred years later, Gudea in his inscriptions detailed the number of workmen (216,000) involved in the project, the cedarwood timbers he had hauled from Lebanon, the other types of timber used for great beams, the "great stones from the mountains, split into blocks"—bitumen from the wells and from the "bitumen lake," copper from the "copper mountains," silver "from its mountain" and "gold from its mountains"; and all the bronze artifacts, and the decorations, and the trimmings, and the stelae and statues. All was described in detail, all was so magnificent and marvelous that, when it was finished, "the Anunnaki were altogether seized with admiration."

The sections in the Gudea inscriptions of the greatest interest are those that deal with the events that preceded the construction of the temple, the determination of its orientation, its equipment and symbolism; we follow primarily the information provided in the inscription known as Cylinder A.

The chain of events, Gudea's record states, began on a certain day, a day of great significance. Referring in the inscriptions to Ninurta by his formal title NIN.GIRSU—"Lord of the Girsu"—this how the record begins:

> *On the day when the fate Heaven-Earth is decreed,*
> *When Lagash lifted its head heavenwards*
> *in accordance with the great ME's,*
> *Enlil cast a favorable eye upon the lord Ningirsu.*

Recording Ninurta's complaint about the delay in the building of the new temple "which is vital to the city in accordance with the ME's," it reports that on that propitious day Enlil finally granted the permission, and he also decreed what the temple's name shall be: "Its king shall name the temple E.NINNU." The edict, Gudea wrote, "made a brilliance in heaven and on Earth."

Having received the permission of Enlil and having obtained the name for the new ziggurat, Ninurta was now free to proceed with the construction. Without losing time, Gudea rushed to supplicate his god to be the one chosen for the task. Offering sacrifices of oxen and kids "he sought the divine will . . . by day and in the middle of the night Gudea lifted up his eyes to his lord Ningirsu; for the command to build his temple he set his eyes." Persisting, Gudea kept praying: "He said and sighed: 'Thus, thus will I speak; thus, thus will I speak; this word I shall bring forth: I am the shepherd, chosen for kingship.' "

Finally the miracle happened. "At midnight," he wrote, "something came to me; its meaning I did not understand." He took his asphalt-lined boat and, sailing on a canal, went to a nearby town to seek an explanation from the oracle goddess Nanshe in her "House of Fate-Solving." Offering prayers and sacrifices that she would solve the riddle of his vision, he proceeded to tell her about the appearance of the god whose command he was to heed:

> *In the dream [I saw]*
> *a man who was bright, shining like Heaven,*
> *great in Heaven, great on Earth,*
> *who by his headdress was a god.*
> *By his side was the Divine Storm Bird;*
> *Like a devouring storm under his feet*
> *two lions crouched, on the right and on the left.*
> *He commanded me to build his temple.*

A celestial omen then occurred whose meaning, Gudea told the oracle goddess, he did not understand: the Sun upon *Kishar*, Jupiter, was suddenly seen on the horizon. A woman then appeared who gave Gudea other celestial directions:

> *A woman—*
> *who was she? Who was she not?*
> *the image of a temple-structure, a ziggurat,*
> *she carried on her head—*
> *in her hand she held a holy stylus,*
> *the tablet of the favorable star of heaven*

she bore,
taking counsel with it.

A third divine being then appeared who had the look of a "hero":

> *A tablet of lapis lazuli his hand held;*
> *the plan of a temple he drew on it.*

And then, before his very eyes, there materialized the signs for construction: "a holy carrying-basket" and a "holy brick mold" in which there was placed "the destined brick."

Having heard the details of the dreamlike vision, the oracle goddess proceeded to tell Gudea what it meant. The first god to appear was Ningirsu (Ninurta); "for thee to build his temple, Eninnu, he commanded." The heliacal rising, she explained, signaled the god Ningishzidda, indicating to him the point of the Sun on the horizon. The goddess was Nisaba; "to build the House in accordance with the Holy Planet she instructed thee." And the third god, Nanshe explained, "Nindub is his name; to thee the plan of the House he gave."

Nanshe then added some instructions of her own, reminding Gudea that the new Eninnu had to provide appropriate places for Ninurta's weapons, for his great aircraft, even for his favorite lyre. Given these explanations and instructions Gudea returned to Lagash and secluded himself in the old temple, trying to figure out what all those instructions meant. "For two days in the sanctuary of the temple he shut himself in, at night he was shut in; the House's plan he contemplated, the vision he repeated to himself."

Most baffling to him, to begin with, was the matter of the new temple's orientation. Stepping up to a high or elevated part of the old temple called *Shugalam*, "the place of the aperture, the place of determining, from which Ningirsu can see the repetition over his lands," Gudea removed some of the "spittle" (mortar? mud?) that obstructed the view, trying to fathom the secrets of the temple's construction; but he was still baffled and perplexed. "Oh my lord Ningirsu," he called out to his god, "Oh son of Enlil: my heart remains unknowing; the meaning is as far from me as the middle of the ocean, as the midst of heaven from me it is distant . . . Oh, son of Enlil, lord Ningirsu—I, I do not know."

He asked for a second omen; and as he was sleeping Ningirsu/Ninurta appeared to him; "While I was sleeping, at my head he stood," Gudea wrote. The god made clear the instructions to Gudea, assuring him of constant divine help:

> *My commands will teach thee the sign*
> *by the divine heavenly planet;*
> *In accordance with the holy rites*
> *my House, the Eninnu,*
> *shall bind Earth with Heaven.*

The god then lists for Gudea all the inner requirements of the new temple, expanding at the same time on his great powers, the awesomeness of his weapons, his memorable deeds (such as the damming of the waters) and the status he was granted by Anu, "the fifty names of lordship, by those ordained." The construction, he tells Gudea, should begin on "the day of the new Moon," when the god will give him the proper omen—a signal: on the evening of the New Year the god's hand shall appear holding a flame, giving off a light "that shall make the night as light as day."

Ninurta/Ningirsu also assures Gudea that he will receive from the very beginning of the planning of the new Eninnu divine help: the god whose epithet was "The Bright Serpent" shall come to help build the Eninnu and its new precinct—"build it to be like the House of the Serpent, as a strong place it shall be built." Ninurta then promises Gudea that the construction of the temple will bring the land abundance: "When my temple-terrace is completed," rains will come on time, the irrigation canals will fill up with water, even the desert "where water has not flowed" shall bloom; there will be abundant crops, and plenty of oil for cooking, and "wool in abundance shall be weighed."

Now "Gudea understood the favorable plan, a plan that was the clear message of his vision-dream; having heard the words of the lord Ningirsu, he bowed his head . . . Now he was greatly wise and understood great things."

Losing no time Gudea proceeded to "purify the city" and organize the people of Lagash, old and young, to form work brigades and enlist themselves in the solemn task. In verses that throw light on the human side of the story, of life and manners and social problems more than four millennia ago, we read that as a way to consecrate themselves for the unique undertaking "the whip of the overseer was prohibited, the mother did not chide her child . . . a maid who had done a great wrong was not struck by her mistress in the face." But the people were asked not only to become angelic; to finance the project, Gudea "levied taxes in the land; as a submission to the lord Ningirsu the taxes were increased" . . .

One can stop here for a moment to look ahead to another construction of a God's Residence, the one built in the wilderness of Sinai for Yahweh. The subject is recorded in detail in the Book of Exodus, beginning in chapter 25. "Speak unto the Children of Israel," Yahweh told Moses, "that they may bring for me a contribution: from every man whose heart shall prompt him thereto shall be taken an allotment for me . . . and they shall build for me a sacred sanctuary, and I shall dwell in their midst. In accordance with all that which I am showing thee, the plan of the Residence and the pattern of all the instruments thereof shall ye make it." Then followed the most detailed architectural instructions—details which make possible the reconstructions of the Residence and its components by modern scholars.

To help Moses carry out these detailed plans, Yahweh decided to provide Moses with two assistants whom Yahweh was to endow with a "divine spirit"—"wisdom and understanding and knowledge of all manner of workmanship." Two men were chosen by Yahweh to be so instructed,

Bezalel and Aholiab, "to carry out all of the sacred work in all of the manner that Yahweh had ordered." These instructions began with the layout plan of the Residence and make clear that it was a rectangular enclosure with its long sides (one hundred cubits) facing precisely south and north and its short sides (fifty cubits in length) facing precisely east and west, creating an east–west axis of orientation (see Fig. 44a).

By now "greatly wise" and "understanding great things," Gudea—to go back to Sumer some seven centuries before the Exodus—launched the execution of the divine instructions in a grand way. By canal and river he sent out boats, "holy ships on which the emblem of Nanshe was raised," to summon assistance from her followers; he sent caravans of cattle and asses to the lands of Inanna, with her emblem of the "star-disk" carried in front; he enlisted the men of Utu, "the god whom he loves." As a result, "Elamites came from Elam, Susians came from Susa; Magan (Egypt) and Melukhah (Nubia) sent a large tribute from their mountains." Cedars were brought from Lebanon, bronze was collected, shiploads of stones arrived. Copper, gold, silver, and marble were obtained.

When all that was ready, it was time to make the bricks of clay. This was no small undertaking, not only because tens of thousands of bricks were needed. The bricks—one of the Sumerian "firsts" which, in a land short of stones, enabled them to build high-rise buildings—were not of the shape and size that we use nowadays: they were usually square, a foot or more on each side and two or three inches thick. They were not identical in all places at all times; they were sometimes just sun-dried, sometimes dried in kilns for durability; they were not always flat, but sometimes concave or convex, as their function required, to withstand structural stress. As is clear from Gudea's as well as other kings' inscriptions, when it came to temples, and even more so to ziggurats, it was the god in charge who determined the size and shape of the bricks; this was such an important step in the construction, and such an honor for the king to mold the first brick, that the kings embedded in the wet bricks a stamped inscription (Fig. 75) with a votive content. This custom, fortunately, made it possible for archaeologists to identify so many of the kings involved in the construction, reconstruction, or repair of the temples.

Gudea devoted many lines in his inscriptions to the subject of the bricks. It was a ceremony attended by several gods and was held on the grounds of the old temple. Gudea prepared himself by spending the night in the sanctuary, then bathing clean and putting on special clothes in the morning. Throughout the land it was a solemn rest day. Gudea offered sacrifices, then went into the old Holy of Holies; there was the brick mold that the god had shown him in the vision-dream and a "holy carrying-basket." Gudea put the basket on his head. A god named Galalim led the procession. The god Ningishzidda held the brick-mold in his hand. He let Gudea pour into the mold water from the temple's copper cup, as a good omen. On a signal from Ninurta, Gudea poured clay into the mold, all the while uttering incantations. Reverently, the inscription says, he carried out the holy rites. The whole city of Lagash "was lying low," waiting for the outcome: will the brick come out right or will it be faulty?

Figure 75

After the sun had shone upon the mold
Gudea broke the mold,
he separated the brick.
The bottom face of the stamped clay he saw;
with a faithful eye he examined it.

The brick was perfect!

He carried the brick to the temple,
the brick raised from the mold.
Like a brilliant diadem he raised it to heaven;
he carried it to the people and raised it.
He put the brick down in the temple;
it was solid and firm.
And the heart of the king
was made as bright as the day.

Ancient, even archaic, Sumerian depictions have been found dealing with the brick ceremony; one of them (Fig. 76) shows a seated deity holding up the Holy Mold, bricks from which are carried to construct a ziggurat.

The time has thus come to start building the temple; and the first step was to mark out its orientation and implant the foundation stone. Gudea wrote that a new place was chosen for the new Eninnu, and archaeologists (see map, Fig. 73) have indeed found its remains on a hill about fifteen hundred feet away from the earlier one, on the mound marked "A" on the excavations map.

We know from these remains that the ziggurat was built

Figure 76

so that its corners would be oriented to the cardinal points; the precise orientation was obtained by first determining true east, then running one or more walls at right angles to each other. This ceremony too was done on an auspicious day for which "the full year" had to come to pass. The day was announced by the goddess Nanshe: "Nanshe, a child of Eridu" (the city of Enki) "commanded the fulfillment of the ascertained oracle." It is our guess that it was the Day of the Equinox.

At midday, "when the Sun came fully forth," the "Lord of the Observers, a Master Builder, stationed at the temple, the direction carefully planned." As the Anunnaki were watching the procedure of determining the orientation "with much admiration," he "laid the foundation stone and marked in the earth the wall's direction." We read later on in the inscription that this Lord of the Observers, the Master Builder, was Ningishzidda; and we know from various depictions (Fig. 77) that it was a deity (recognized by his horned cap) who implanted the conical cornerstone on such occasions.

Apart from depictions of the ceremony, showing a god with the horned headdress implanting the conical "stone," such representations cast in bronze suggest that the "stone" was actually a bronze one; the use of the term "stone" is not unusual, since all metals resulting from quarrying and mining were named with the prefix NA, meaning "stone" or "that which is mined." In this regard it is noteworthy that in the Bible the laying of the corner or First Stone was also considered a divine or divinely inspired act signifying the Lord's blessing to the new House. In the prophecy by Zechariah about the rebuilding of the Temple in Jerusalem, he relates how Yahweh showed him in a vision "a man holding a measuring cord in his hand," and how he was told that this divine emissary would come to measure the four sides of a rebuilt and greater Jerusalem with its new Lord's House, whose stones shall rise sevenfold after the Lord will place for him the First Stone. "And when they shall see the bronze stone in the hands of Zerubbabel" (the one chosen by Yahweh to rebuild the Temple) all nations

Figure 77

will know that it was the Lord's will. On that occasion too the men chosen to carry out the temple's rebuilding were named by Yahweh.

In Lagash, once the cornerstone was embedded by the god Ningishzidda, Gudea was able to lay the temple's foundations, by now "like Nisaba knowing the meaning of numbers."

The ziggurat built by Gudea, scholars have concluded, was one of seven stages. Accordingly, seven blessings were pronounced as soon as the laying of the foundation stone was completed and the temple's orientation set and Gudea began to place the bricks along the marking on the ground:

> *May the bricks rest peacefully!*
> *May the House by its plan rise high!*
> *May the divine Black Storm Bird*

be as a young eagle!
May it be like a young lion awesome!
May the House have the brilliance of Heaven!
May joy abound at the prescribed sacrifices!
May Eninnu be a light unto the world!

Then did Gudea begin to build the "House, a dwelling he established for his lord Ningirsu . . . a temple truly a Heaven-Earth mountain, its head reaching heavenward . . . Joyfully did Gudea erect the Eninnu with Sumer's firm bricks; the great temple he thus constructed."

With no stones to be quarried in Mesopotamia, the "Land between the rivers" which was covered with an avalanche of mud during the Deluge, the only building materials were the mud or clay bricks, and all the temples and ziggurats were so built. The statement by Gudea that the Eninnu was erected "with Sumer's firm bricks" is thus a mere statement of fact. What is puzzling is the detailed list by Gudea of other materials used in the construction. We refer here not to the various woods and timbers, which were commonly used in temple constructions, but to the variety of metals and stones employed in the project—materials all of which had to be imported from afar.

The king, we read in the inscriptions, "the Righteous Shepherd," "built the temple bright with metal" bringing copper, gold, and silver from distant lands. "He built the Eninnu with stone, he made it bright with jewels; with copper mixed with tin he held it fast." This is undoubtedly a reference to bronze which, in addition to its use for various listed artifacts, apparently was also used to clamp together stone blocks and metals. The making of bronze, a complex process involving the mixing of copper and tin under great heat in specified proportions, was quite an art; and indeed Gudea's inscription makes it clear that for the purpose a *Sangu Simug,* a "priestly smith," working for the god Nintud, was brought over from the "Land of smelting." This priestly smith, the inscription adds, "worked on the temple's facade; with two handbreadths of bright stone he faced over the brickwork; with diorite and one handbreadth of bright stone he . . . " (the inscription is too damaged here to be legible).

Not just the mere quantity of stones used in the Eninnu but the outright statement that the brickwork was faced with bright stone of a certain thickness—a statement that until now has not drawn the attention of scholars—is nothing short of sensational. We know of no other instance of Sumerian records of temple construction that mention the facing or "casing" of brickwork with stones. Such inscriptions speak only of brickwork—its erection, its crumbling, its replacement—but *never of a stone facing* over the brick facade.

Incredibly—but as we shall show, not inexplicably—the facing of the new Eninnu with bright stones, unique in Sumer, emulated the *Egyptian* method of facing step-pyramids with bright stone casings to give them smooth sides!

The Egyptian pyramids that were built by pharaohs began with one built by King Zoser at Sakkara (south of Memphis) circa 2650 B.C. (Fig. 78). Rising in six steps within a rectangular sacred precinct, it was originally faced with bright limestone casing stones of which only traces now remain: its casing stones, as those of ensuing pyramids, were removed by later rulers to be used in their own monuments.

The Egyptian pyramids, as we have shown and proved in *The Stairway to Heaven,* began with those built by the Anunnaki themselves—the Great Pyramid and its two companions at Giza. It was they who devised the casing with bright stones of what were in their core step-pyramids, giving them their renowned smooth sides. That the new Eninnu in Lagash, commissioned by Ninurta at about the same time

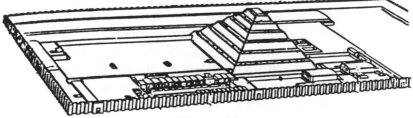

Figure 78

as Stonehenge became truly a *stone*-henge, emulated an Egyptian pyramid's stone facing, is a major clue for the resolution of the Stonehenge enigma.

Such an unexpected link to ancient Egypt, as we have been showing, was only one among many. Gudea himself was alluding to these connections when he stated that the shape of the Eninnu and its casing with bright stones were based on information provided by Nisaba "who was taught the plan of the temple by Enki" in the "House of Learning." That academy was undoubtedly in one of Enki's centers; and Egypt, it will be recalled, was the domain allotted to Enki and his descendants when Earth was divided.

The Eninnu project involved the participation of quite a number of gods; Nisaba, who had appeared to Gudea in the first vision with the star map, was not the only female among them. Let us look at the full list, then highlight the female roles.

First there was Enlil, who began the process by granting the permission to Ninurta to build the new temple. Then Ninurta appeared to Gudea, informing him of the divine decision and of his (Gudea's) selection to be the builder. In his vision Ningishzidda indicated to him the celestial point where the Sun rose, Nisaba pointed with a stylus to the favorable star, and Nindub drew the plan of the temple on a tablet. In order to understand all that, he consulted Nanshe, the oracle goddess. Inanna/Ishtar and Utu/Shamash enlisted their followers in obtaining the rare building materials. Ningishzidda, with the participation of a god named Galalim, was involved in molding the bricks. Nanshe chose the auspicious day on which to start the construction. Ningishzidda then determined the orientation and laid the cornerstone. Before the Eninnu was declared fit for its purpose, Utu/Shamash examined its alignment with the Sun. The individual shrines built alongside the ziggurat honored Anu, Enlil, and Enki. And the final purification and consecration rites, before Ninurta/Ningirsu and his spouse Bau moved in, involved the deities Ninmada, Enki, Nindub, and Nanshe.

Astronomy clearly played a key role in the Eninnu project; and two of the deities involved, Nanshe and Nisaba, were female astronomer-gods. They applied their specialized knowledge of astronomy, mathematics, and metrology not only to temple construction (as in Gudea's case), but also to general productive purposes as well as in ritual roles. One, however, was trained in the academy of Eridu; the other in that of Nippur.

Nanshe, who identified for Gudea the celestial role of each of the deities that appeared to him in his vision and determined the precise calendrical day (of the equinox) for orienting the temple, is called in the Gudea inscriptions "a daughter of Eridu" (Enki's city in Sumer). Indeed, in the major God Lists of Mesopotamia, she was called NIN.A— "Lady of Water"—and shown as a daughter of Ea/Enki. The planning of waterways and the locating of fountainheads were her specialty; her celestial counterpart was the constellation Scorpio—*mul GIR.TAB* in Sumerian. The knowledge she contributed to the building of the Eninnu in Lagash was thus that of the Enki'ite academies.

A hymn to Nanshe in her role as determiner of the New Year Day has her sitting in judgment on Mankind on that day, accompanied by Nisaba in the role of Divine Accountant who tallies and measures the sins of those who are judged, such as the sin of he "who substituted a small weight for a large weight, a small measure for a large measure." But while the two goddesses were frequently mentioned together, Nisaba (some scholars read her name Nidaba) was clearly listed among the Enlilites, and was sometimes identified as a half sister of Ninurta/Ningirsu. Although she was in later times deemed to be a goddess who blesses the crops—perhaps because of her association with the calendar and weather—she was described in Sumerian literature as one who "opens men's ears," i.e. teaches them wisdom. In one of several School Essays compiled by Samuel N. Kramer (*The Sumerians*) from scattered fragments, the *Ummia* ("Word-knower") names Nisaba as the patron goddess of the E.DUB.BA ("House of Inscribed Tablets"), Sumer's principal academy for scribal arts. Kramer called her "the Sumerian goddess of Wisdom."

Nisaba was, in the words of D.O. Edzard (*Götter und Mythen im Vorderen Orient*), the Sumerian goddess of "the art of writing, mathematics, science, architecture and astronomy." Gudea specifically described her as the "goddess who knows numbers"—a female "Einstein" of antiquity . . .

The emblem of Nisaba was the Holy Stylus. A short hymn to Nisaba on a tablet unearthed in the ruins of the sacred precinct of Lagash (Fig. 79) describes her as "she who acquired fifty great ME's" and as possessor of the "stylus of seven numbers." Both numbers were associated with Enlil and Ninurta: the numerical rank of both was fifty, and one of Enlil's epithets (as commander of Earth, the seventh planet) was "Lord of Seven."

With her Holy Stylus Nisaba pointed out to Gudea the "favorite star" on the "star tablet" that she held on her knees; the implication is that the star tablet had drawn on it more than one star, so that the correct one for the orientation had to be pointed out from among several stars. This conclusion is strengthened by the statement in *The Blessing of Nisaba by Enki* that Enki had given her as part

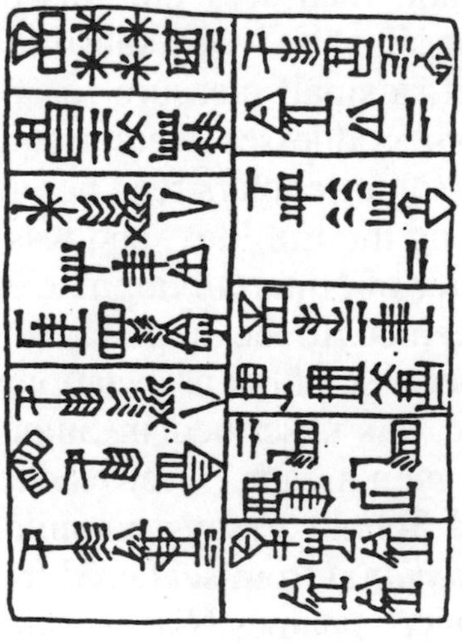

Figure 79

of her schooling "the holy tablet of the heavenly stars"—again "stars" in the plural.

The term MUL in Sumerian (*Kakkab* in Akkadian), meaning "celestial body," was applied to both planets and stars, and one wonders what heavenly bodies were shown on the star map possessed by Nisaba, whether they were stars or planets or (probably) both. The opening line of the text shown in Fig. 79, paying homage to Nisaba as a great astronomer, calls her NIN MUL.MUL.LA—"Lady of Many Stars." The intriguing aspect of this formulation is that the term "many stars" is written not with a star sign together with the determinative sign for "many," but with four star signs. The only plausible explanation for this unusual formulation is that Nisaba could point out, on her sky map, the four stars that we continue to use for determining the cardinal points.

Her great wisdom and scientific knowledge were expressed in Sumerian hymns by the statement that she was "perfected with the fifty great ME's"—those enigmatic "divine formulas" that, like computer disks, were small enough to be carried by hand though each contained a vast amount of information. Inanna/Ishtar, a Sumerian text related, went to Eridu and tricked Enki into giving her one hundred of them. Nisaba, on the other hand, did not have to steal the fifty ME's. A poetic text compiled from fragments and rendered into English by William W. Hallo (in a lecture titled "The Cultic Setting of Sumerian Poetry") that he called *The Blessing of Nisaba by Enki,* makes clear that in addition to her Enlilite schooling Nisaba was also a graduate of the Eridu academy of Enki. Extolling Nisaba as "Chief scribe of heaven, record-keeper of Enlil, all-knowing sage of the gods" and exalting Enki, "the craftsman of Eridu" and his "House of Learning," the hymn says of Enki:

He verily opened the House of Learning for Nisaba;
He verily placed the lapis lazuli tablet on her knee,
to take counsel with the holy tablet of the heavenly stars.

The "cult city" of Nisaba was called Eresh ("Foremost Abode"); its remains or location were never discovered in Mesopotamia. The fifth stanza of this poem suggests that it was located in the "Lower World" (*Abzu*) of Africa, where Enki oversaw the mining and metallurgical operations and conducted his experiments in genetics. Listing the various distant locations where Nisaba was also schooled under Enki's aegis, the poem states:

> *Eresh he constructed for her,*
> *in abundance created of pure little bricks.*
> *She is granted wisdom of the highest degree*
> *in the Abzu, great place of Eridu's crown.*

A cousin of Nisaba, the goddess ERESH.KI.GAL ("Foremost Abode in the Great Place"), was in charge of a scientific station in southern Africa and there shared control of a Tablet of Wisdom with Nergal, a son of Enki, as a marriage dowry. It is quite possible that it was there that Nisaba acquired her additional schooling.

This analysis of Nisaba's attributes can help us to identify the deity—let us call her Goddess of Astronomers—appearing on an Assyrian tablet (Fig. 80). She is shown inside

Figure 80

a gateway surmounted by the stepped viewing positions. She holds a pole-mounted viewing instrument, identified here by the crescent as one for viewing the Moon's movements, i.e. for calendrical purposes. And she is further identified by the four stars—the symbol, we believe, of Nisaba.

One of the oddest statements made by Gudea when he described the deities who appeared to him concerned Nisaba: "The image of a temple-structure, a ziggurat, she carried on her head." The headdress of Mesopotamian deities was distinguished by its pairs of horns; that gods or goddesses would instead wear on their heads the image of a temple or an object was absolutely unheard of. Yet, in his inscription, that is how Gudea described Nisaba.

He was not imagining things. If we examine illustration 80, we will see that Nisaba is indeed carrying on her head the image of a temple-ziggurat, just as Gudea had stated. But it is not a stepped structure; rather, it is the image of a smooth-sided pyramid—an *Egyptian* pyramid!

Moreover, not only is the ziggurat Egyptianized—the very custom of wearing such an image on the head is Egyptian, especially as it applied to Egyptian goddesses. Foremost of them were Isis, the sister-wife of Osiris (Fig. 81a) and Nephtys, their sister (Fig. 81b).

Was Nisaba, an Enlilite goddess schooled in Enki's academy, Egyptianized enough to be wearing this kind of headgear? As we pursue this investigation, many similarities between Nisaba and Sesheta, the female assistant of Thoth in Egypt, come to light. In addition to the attributes and function of Sesheta that we have already reviewed, there were others that closely matched those of Nisaba. They included her role as "the goddess of the arts of writing and of science," in the words of Hermann Kees (*Der Götterglaube in Alten Aegypten*). Nisaba possessed the "stylus of seven numbers"; Sesheta too was associated with the number seven. One of her epithets was "Sesheta means seven" and her name was often written hieroglyphically by the sign for seven placed above a bow. Like Nisaba, who had appeared to Gudea with the image of a temple-structure on

Figures 81a and 81b

her head, so was Sesheta depicted with the image of a twin-
towered structure on her head, above her identifying star-
and-bow symbol (Fig. 82). She was a "daughter of the
sky," a chronologer and chronographer; and like Nisaba,
she determined the required astronomical data for the royal-
temple builders.

According to the Sumerian texts, the consort of Nisaba
was a god called *Haia*. Hardly anything is known of him,
except that in the judgment procedures on New Year's Day
supervised by Nanshe, he was also present, acting as the
balancer of the scales. In Egyptian beliefs Judgment Day
for the pharaoh was when he died, at which time his heart
was weighed to determine his fate in the Afterlife. In Egyptian theology, the god who balanced the scales was Thoth,
the god of science, astronomy, the calendar, and of writing
and record keeping.

Such an overlapping of identities between the deities who
provided the astronomical and calendrical knowledge for
the Eninnu reveals an otherwise unknown state of cooperation between the Sumerian and Egyptian Divine Architects
in Gudea's time.

It was, in many respects, an unusual phenomenon; it
found expression in the unique shape and appearance of the

Figure 82

Eninnu and in the establishment within its sacred precinct of an extraordinary astronomical facility. It all involved and revolved around the calendar—the gift to Mankind by the divine Keepers of the Secrets.

After the construction of the Eninnu ziggurat was completed, much effort and artistry went into its adorning, not only outside but also inside; portions, we learn, of the "inner shrine" were overlaid with "cedar panels, attractive to the eye." Outside, rare trees and bushes were planted to create a pleasant garden. A pool was built and filled with rare fish—another unusual feature in Sumerian temple precincts and one which is akin to Egyptian ones, where a sacred pool was a common feature.

"The dream," Gudea wrote, "was fulfilled." The Eninnu was completed, "like a bright mass it stands, a radiant brightness of its facing covers everything; like a mountain which glows it joyously rises.''

Now he turned his attention and efforts to the Girsu, the sacred precinct as such. A ravine, ''a great dump'' was filled up: ''with the wisdom granted by Enki divinely he did the grading, enlarging the area of the temple terrace.'' Cylinder A alone lists more than fifty separate shrines and temples built adjoining the ziggurat to honor the various gods involved in the project as well as Anu, Enlil, and Enki. There were enclosures, service buildings, courts, altars, gates; residences for the various priests; and, of course, the special dwelling and sleeping quarters of Ningirsu/Ninurta and his spouse Bau.

There were also special enclosures or facilities for housing the Divine Black Bird, the aircraft of Ninurta, and for his awesome weapons; as well as places at which the astronomical-calendrical functions of the new Eninnu were to be performed. There was a special place for ''the Master of Secrets,'' and the new *Shugalam,* the high place of the aperture, the ''place of determining whose awesomeness is great, where the Brilliance is announced.'' And there were two buildings connected with the ''solving of the cords'' and the ''binding with the cords'' respectively— facilities whose purpose has eluded scholars but which had to be connected with celestial observations, for they were located next to, or were part of, the structures called ''Uppermost Chamber'' and ''Chamber of the seven zones.''

There were certain other features that were added to the new Eninnu and its sacred precinct that indeed made it as unique as Gudea had boasted; we shall discuss them in the detail they deserve further on. There was also a need, as the text makes clear, to await a certain specific day—New Year's Day, to be precise—before Ninurta and his spouse Bau could actually move into the new Eninnu and make it their dwelling abode.

Whereas Cylinder A was devoted to the events leading to the construction of the Eninnu and the construction itself, Gudea's inscriptions on Cylinder B deal with the rites connected with the consecration of the new ziggurat and its sacred precinct and the ceremonies involved in the actual arrival of Ninurta and Bau in the *Girsu*—reaffirming his title as NIN.GIRSU, "Lord of Girsu"—and their entry into their new dwelling place. The astronomical and calendrical aspects of these rites and ceremonies enhance the data with which the Cylinder A inscriptions are filled.

While the arrival of the inauguration day was awaited—for the better part of a year—Gudea engaged in daily prayers, the pouring of libations, and the filling up of the new temple's granaries with food from the fields and its cattle pens with sheep from the pastures. Finally the designated day arrived:

> *The year went round,*
> *the months were completed;*
> *the New Year came in the heavens—*
> *the "Month of the Temple" began.*

On that day, as the "new Moon was born," the dedication ceremonies began. The gods themselves performed the purification and consecration rites: "Ninmada performed the purification; Enki granted a special oracle; Nindub spread incense; Nanshe, the Mistress of Oracles, sang holy hymns; they consecrated the Eninnu, made it holy."

The third day, Gudea recorded, was a bright day. It was on that day that Ninurta stepped out—"with a bright radiance he shone." As he entered the new sacred precinct, "the goddess Bau was advancing on his left side." Gudea "sprinkled the ground with an abundance of oil . . . he brought forth honey, butter, wine, milk, grain, olive oil . . . dates and grapes he piled up in a heap—food untouched by fire, food for the eating by the gods."

The entertainment of the divine couple and the other gods with fruits and other uncooked foods went on until midday. "When the Sun rose high over the country" Gudea "slaughtered a fat ox and a fat sheep" and a feast of roasted meats with much wine began; "white bread and milk they brought by day and through the night"; and "Ninurta, the warrior

of Enlil, taking food and beer for drink, was satisfied.'' All
the while Gudea ''made the whole city kneel, he made the
whole country prostrate itself . . . By day there were peti-
tions, by night prayers.''

''At the morning aurora''—at dawn—''Ningirsu, the
warrior, entered the Temple; into the Temple its lord came;
giving a shout like the cry of battle, Ningirsu advanced into
his temple.'' ''It was,'' observed Gudea, ''like the rising
of the Sun over the land of Lagash . . . and the Land of
Lagash rejoiced.'' It was also the day on which the harvest
began:

> *On that day,*
> *when the Righteous God entered,*
> *Gudea, on that day,*
> *began to harvest the fields.*

Following a decree of Ninurta and the goddess Nanshe,
there followed seven days of repentance and atonement in
the land. ''For seven days the maid and her mistress were
equal, master and slave walked side by side . . . of the evil
tongue the word was changed to good . . . the rich man did
not wrong the orphan, no man oppressed the widow . . . the
city restrained wickedness.'' At the end of the seven days,
on the tenth day of the month, Gudea entered the new temple
and for the first time performed there the rites of the High
Priest, ''lighting the fire in the temple-terrace before the
bright heavens.''

A depiction on a cylinder seal from the second millennium
B.C., found at Ashur, may well have preserved for us the
scene that had taken place a thousand years earlier in Lagash:
it shows a High Priest (who as often as not was also the
king, as in the case of Gudea) lighting a fire on an altar as
he faces the god's ziggurat, while the ''favorite planet'' is
seen in the heavens (Fig. 83).

On the altar, ''before the bright heavens, the fire on the
temple-terrace increased.'' Gudea ''oxen and kids sacrificed
in numbers.'' From a lead bowl he poured a libation. ''For
the city below the temple he pleaded.'' He swore everlasting

Figure 83

allegiance to Ningirsu, "by the bricks of Eninnu he swore, a favorable oath he swore."

And the god Ninurta, promising Lagash and its people abundance, that "the land may bear whatever is good," to Gudea himself said: "Life shall be prolonged for thee."

Appropriately, the Cylinder B inscription concludes thus:

House, rising heavenward as a great mountain,
its luster powerfully falls on the land
as Anu and Enlil the fate of Lagash determine.

Eninnu, for Heaven-Earth constructed,
the lordship of Ningirsu
to all the lands it makes known.

O Ningirsu, thou art honored!
The House of Ningirsu is built;
Glory be unto it!

7

A STONEHENGE
ON THE EUPHRATES

There is a wealth of information in the inscriptions of Gudea; the more we study them and the special features of the Eninnu he built, the more astounded we shall be.

Perusing the texts verse by verse and visualizing the great new temple-terrace and its ziggurat, we shall discover amazing celestial features of the "Bond Heaven-Earth"; one of the earliest if not the very earliest association of a temple with the zodiac; the appearance of sphinxes in Sumer at a totally unexpected time; an array of links with Egypt and especially with one of its gods; and a "mini-Stonehenge" in the Land Between the Rivers . . .

Let us begin with the first task Gudea undertook after the construction of the ziggurat was completed and the temple-terrace formed. It was the erection of seven upright stone pillars at seven carefully selected positions. Gudea, the inscription states, made sure that they be firmly erected: he "laid them on a foundation, on bases he erected them."

The stelae (as scholars call these upright stones) must have been of great importance, for Gudea spent a full year in bringing the rough stone blocks, from which the uprights were carved to shape, from a distant source to Lagash; and another year to cut and shape them. But then, in a frenzied effort that lasted a precise seven days during which the work was carried out without stopping, without rest, the seven stelae were set up in their proper places. If, as the information given suggests, the seven stelae were positioned in some astronomical alignment, then the speed becomes understandable, for the longer the setting up would have taken,

170

the more misaligned the celestial bodies would have become.

Signifying the importance of the stelae and their position is the fact that Gudea gave each one a ''name'' made up of a long sacred utterance evidently related to the position of the stela (e.g. ''on the lofty terrace,'' facing the ''gate of the river-bank'' or another one ''opposite the shrine of Anu''). Although the inscription stated unequivocally (column XXIX line 1) that ''seven stelae were erected'' in those seven hectic days, the names of only six locations are given. In respect to one, presumably the seventh stela, the inscription states that it ''was erected toward the rising sun.'' Since by then all the required orientations of the Eninnu had already been fixed, starting with the divine instructions and the laying of the cornerstone by Ningishzidda, neither the six spread out stelae nor the seventh ''erected toward the rising sun'' were required for orienting the temple. Another, different purpose had to be the motive; the only logical conclusion is that it involved observations other than determining the Day of the Equinox (i.e. of the New Year)— some astronomical-calendrical observations of an unusual nature, justifying the great effort in obtaining and shaping the stelae and the haste in setting them up.

The enigma of these erected stone pillars begins with the question, why so many when two are enough to create a line of sight, say toward the rising Sun. The puzzle is engulfed by incredulity when we read on in the inscription the sensational statement that the six whose locations were named were placed by Gudea ''*in a circle.*'' Did Gudea use the stelae to form a stone *henge*—in ancient Sumer, more than five thousand years ago?

Gudea's inscription indicates, according to A. Falkenstein (*Die Inschriften Gudeas von Lagash*), the existence of an avenue or pathway which—as at Stonehenge!—could provide an unimpeded sightline. He noted that the stela which was ''toward the rising Sun'' stood at one end of a pathway or avenue called ''Way to the high position.'' At the other end of this way was the *Shugalam,* the ''High place whose awesomeness is great, where the Brilliance is raised.'' The term SHU.GALAM meant, according to Falkenstein, ''Where the hand is raised''—a high place from which a signal is given. Indeed, the Cylinder A inscription asserts that ''At the radiant entrance of Shu-galam, Gudea stationed a favorable image; toward the rising Sun, in the destined place, the emblem of the Sun he established.''

We have already discussed the functions of the *Shugalam* when Gudea had gone to it, in the old temple, to remove the mortar or mud that obstructed the view through it. It was, we found, ''the place of the aperture, the place of determining.'' There, the inscription stated, ''Ninurta could see the repetitions''—the annual celestial cycle—''over his lands.'' The description brings to mind the ceiling aperture about which there was so much arguing on Mount Zaphon between Ba'al and the divine architect who came from Egypt to design the new temple in Lebanon.

Some additional light on the enigmatic purpose of such a skylight or aperture in the ceiling can be obtained from the examination of the Hebrew term for such a contraption and its Akkadian roots. It is *Tzohar* and appears just once in the Bible to describe the only aperture in the ceiling of the otherwise hermetically sealed Noah's Ark. The meaning, all agree, is ''a ceiling window through which a beam of light can shine in.'' In modern Hebrew the term is also used to denote ''zenith,'' the point in the sky directly over-head; and both in modern Hebrew and biblical texts the term *Tzohora' im* that derived from it meant and still means ''mid-day,'' when the Sun is directly overhead. *Tzohar* was thus not just a simple aperture, but one intended to let a beam of the Sun shine into a darkened enclosure at a certain time of the day; spelled slightly differently, *Zohar,* the term acquired the meaning ''brightness, brilliance.'' All stem from the Akkadian, the mother tongue of all the Semitic languages, in which the words *tzirru, tzurru* meant ''lighten up, shine'' and ''be high.''

At the *Shugalam,* Gudea wrote, he ''fixed the image of the Sun.'' All the evidence suggests that it was a viewing device through which the rising Sun—undoubtedly on Equi-

nox Day, to judge from all the data in the inscriptions—
was observed to determine and announce the arrival of the
New Year.

Was the concept underlying the structural arrangement
the same as (possibly) the one on Mount Zaphon and (certainly) as at the Egyptian temples, where a beam of sunlight
passed along the preselected axis to light up the Holy of
Holies at sunrise on the prescribed day?

In Egypt the Sun Temples were flanked by two obelisks
(Fig. 84) which the pharaohs erected so that they might
be granted long life; their function was to guide the Sun's
beam on the prescribed day. E.A. Wallis Budge (*The
Egyptian Obelisk*) pointed out that the pharaohs, such as
Ramses II and Queen Hatshepsut, always set up these
obelisks in pairs. Queen Hatshepsut even wrote her royal
name (within a cartouche) between two obelisks (Fig. 85a)
to imply that the Blessed Beam of Ra shone on her on
the crucial day.

Scholars have noted that Solomon's Temple also had two
pillars erected at its entrance (Fig. 85c); like the uprights
at the Eninnu which were given names by Gudea, so were
the two pillars named by Solomon:

> *And he set up the pillars*
> *in the porch of the temple.*
> *He set up the right pillar*

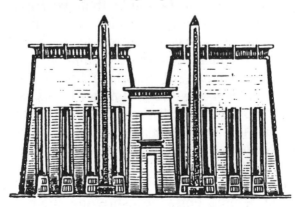

Figure 84

and called the name thereof Yakhin;
and he set up the left pillar
and called the name thereof Bo'az.

While the meaning of the two names eludes scholars (the best assumption is "Yahweh makes firm" and "In him is strength"), the shape, height, and makeup of the pillars is described in the Bible (mainly I Kings chapter 7) in detail. The two pillars were made of cast bronze, eighteen cubits (some twenty-seven feet) high. Each pillar supported a complex "headband" around which, as a crown, there was placed a corolla whose serrated top created seven protrusions; one of them (or both, depending on the way the verse is read) was "encircled by a cord twelve cubits long." (Twelve and seven are the predominant numbers in the Temple.)

The Bible does not state the purpose of these pillars, and theories have ranged from purely decorative or symbolic to a function akin to that of the pair of obelisks that flanked the entrances to the temples in Egypt. In this regard a clue is suggested by the Egyptian word for "obelisk," which was *Tekhen*. The term, Budge wrote, "was a very old word, and we find it in the dual in the Pyramid Texts which were written before the close of the VIth Dynasty." As to the meaning of the word, which he did not know, he added: "The exact meaning of *Tekhen* is unknown to us and it is probable that the Egyptians had forgotten it at a very early period." This raises the possibility that the word was a foreign term, a "loanword" from another language or country, and we on our part believe that the source, of both the biblical *Yakhin* and the Egyptian *Tekhen* was the Akkadian root *khunnu* which meant "to establish correctly" as well as "to start a light" (or fire). The Akkadian term may even be traced back to the earlier Sumerian term GUNNU which combined the meanings "daylight" with "tube, pipe."

These linguistic clues sit well with earlier Sumerian depictions of temple entrances showing them flanked by pillars to which circular devices were attached (Fig. 85b).

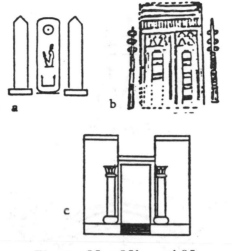

Figures 85a, 85b, and 85c

These must have been the forerunners of all such pairs of uprights, pillars, or obelisks elsewhere, for they appear on the Sumerian depictions millennia before the others. The search for answers to the puzzle of these uprights is further assisted by examining the term used by Gudea in his inscriptions to describe the stone uprights. He called all seven of them NE.RU— from which the Hebrew word *Ner,* meaning "candle," stems. Sumerian script evolved by the scribe's making wedgelike markings with a stylus on wet clay to emulate the original drawing of the object or action for which the sign stood. We find that the original pictograph for the term *Neru* was that of two—two, not one—pillars set upon stable bases with antennalike protrusions (Fig. 86).

Such paired pillars, guiding (actually or symbolically) the Sun's beam on a specific day were sufficient if only one solar position—equinoctial *or* solstitial—was involved. If such a single determination was intended at the Girsu, two stelae, in alignment with the Shugalam, would have sufficed. But Gudea set up seven of them, six in a circle and the seventh in alignment with the Sun. To form a line of sight, this odd pillar could have been positioned either in the circle's center, or outside of it in the avenue. Either

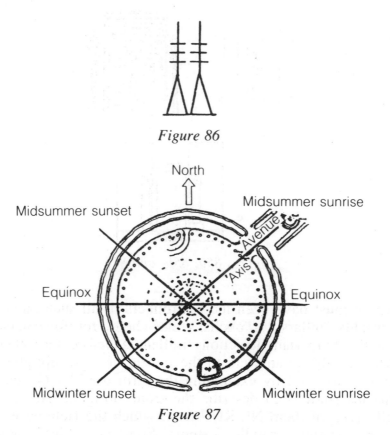

Figure 86

Figure 87

way, the outcome would indicate uncanny similarities to
Stonehenge in the British Isles.

Six outer or circumference points with one in the center
would have created a layout (Fig. 87) that, as in Stonehenge
II—belonging to the same time—provided alignments not
only with the equinoxes but also with the four solstice points
(midsummer sunrise and sunset, midwinter sunrise and sunset). Since the Mesopotamian New Year was firmly anchored to the equinoxes, resulting in ziggurats whose
determining corner was oriented to the east, an arrangement
of stone pillars that incorporated fixings of the solstices was
a major innovation. It also indicated a decisive ''Egyptian''
influence, for it was at Egyptian temples that an orientation
linked to the solstices was the dominant feature—certainly
by Gudea's time.

If, as Falkenstein's study suggests, the seventh pillar was not within the circle of six stelae but outside of it—in the pathway or avenue leading to the Shugalam, an even more astounding similarity emerges, not to the later Stonehenge but to the earliest one, to Stonehenge I, where—we may recall—there were only seven stones: the four Station Stones forming a rectangle, two Gateway Stones that flanked the beginning of the Avenue, and the Heel Stone that marked out the sightline—an arrangement of seven stone uprights illustrated in Fig. 88. Since at Stonehenge the Aubrey Holes were part of phase I, the sightline could be easily determined by a viewer at hole 28 directing his gaze through a post inserted in hole 56, watching for the Sun to appear above the Heel Stone on the propitious day.

Such a similarity in layouts would be even more significant than the first alternative, for—as we have reported earlier—the rectangle formed by the four Station Stones implied lunar observations in addition to the solar ones. The realization of this rectangular arrangement led both Newham

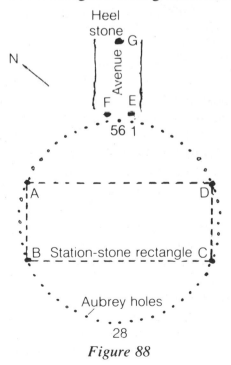

Figure 88

and Hawkins to far-reaching conclusions regarding the sophistication of the planners of Stonehenge I. But since Stonehenge I preceded the Eninnu by about seven centuries, the similarity would have to imply that whoever had planned the layout of the seven uprights in the Eninnu copied from whoever had planned Stonehenge I.

Such a kinship between the two structures, in two different parts of the world, seems incredible; it will, however, become credible as we bring to light more amazing aspects of Gudea's Eninnu.

The six-plus-one circle just described was not the only stone circle on the platform of the new Eninnu.

Boasting that he accomplished "great things" that called for unusual "wisdom" (scientific knowledge), Gudea proceeded to describe, after the section dealing with the stelae, the "crownlike circle for the new Moon"—a creation in stone so unique that "its name in the midst of the world he caused to brightly go forth." This second circle was arranged as a "round crown for the new Moon" and consisted of thirteen stones erected "like heroes in a network"—a most figurative way, it seems to us, to describe *a circle of upright stones connected at the top by lintels to form a "network" akin to the Trilithons at Stonehenge!*

While the possibility that the first smaller circle served lunar as well as solar functions can be only surmised, the second larger circle was undoubtedly intended to observe the Moon. Judging by the repeated references in the inscriptions to the New Moon, the lunar observations were geared to the Moon's monthly cycle, its waxing and waning in the course of four quarters. Our interpretation of the crownlike circle is reinforced by the statement that this circle consisted of two groups of megaliths—one of six and the other of seven, with the latter apparently more elevated or taller than the first.

At first glance the arrangement of *thirteen* (six plus seven) megaliths, connected at their tops by lintels to form a "crown," seems to be an error, because we expect to find only twelve pillars (which in a circle create twelve apertures) if the arrangement is related to the twelve months of the lunar phases. The presence of thirteen pillars, however, does make sense if account were taken of the need to add one month every now and then for intercalation purposes. If so, the amazing stone circles in the Girsu were also the first instance where calendars made of stone meshed to correlate the solar and lunar cycles.

(One wonders whether these stone circles in the Girsu somehow presaged the introduction of the seven-day week—a division of time whose origin has evaded scholars—the biblical week which totaled seven by adding the six days of creation to the final additional day of rest. The number seven appears twice, in the first arrangement of pillars and as part of the second circle; and it is quite possible that somehow days were counted according to either group, leading to a repetition of periods of seven days. Also, four phases of the Moon multiplied by the thirteen pillars would divide the year into fifty-two weeks of seven days each).

Whatever the astronomical-calendrical possibilities inherent in the two circles (and we have probably only touched upon the very basic ones), *it is evident that in the Girsu of Lagash a solar-lunar stone computer was put into operation.*

If all this begins to sound like a *"Stonehenge on the Euphrates"*—a mini-Stonehenge erected by a Sumerian king in the *Girsu* of Lagash at about the same time that Stonehenge in the British Isles became a truly stone circle circa 2100 B.C.—there is more to come. It was at that time that the second type of stone, the bluestones, was brought to the plain of Salisbury from a distant source. This too enhances the similarities: Gudea too hauled not one but two types of stones from a great distance, "from the stone mountains" of Magan (Egypt) and Melukhah (Nubia), both in Africa. We read in the inscription on Cylinder A that it took a full year to obtain these stone blocks from "stone mountains which no [Sumerian] king had entered before." To reach them, Gudea "into the mountains made a way, and their great stones he brought out in blocks; shiploads of *Hua* stones and *Lua* stones.''

Though the meaning of the names of the two types of stones remains undeciphered, their distant origin is clearly stated. Coming from two African sources, they were first transported by land via a new way made by Gudea, then carried by ships over sea routes to Lagash (which was connected to the Euphrates River by a navigable canal).

As at Salisbury Plain in the British Isles so was it in the Mesopotamian plain: stones hauled from afar, stones especially selected, set up in two circles. As at Stonehenge I, seven pillars played a key role; as in all the phases of Stonehenge in Lagash, too, a large megalith created the desired sightline toward the principal solar orientation. In both places a stone ''computer'' was created to serve as a solar-lunar observatory.

Were both, then, created by the same scientific genius, by the same Divine Architect—or were they simply the result of accumulated scientific traditions that found expression in similar structures?

While general scientific knowledge as applied to astronomy and the calendar undoubtedly played a role, the hand of a specific Divine Architect cannot be ignored. In earlier chapters we have pointed out the key difference in design between Stonehenge and all the other temples of the Old World: the former was based on circular formations to observe the heavens; the latter were all built with right angles (rectangular or square). This difference is evident not only in the general plan of the other temples but also in the several instances where stone uprights were found, emplaced in a pattern suggesting an astronomical-calendrical function. An outstanding example was found at Byblos, on a promontory overlooking the Mediterranean Sea. The Holy of Holies of its temple, square in shape, was flanked by upright stone monoliths. They were set up in alignments suggesting observations of equinoxes and solstices; but none were arranged in a circle. So apparently was the case at a Canaanite site, Gezer, near Jerusalem, where the discovery of a tablet inscribed with the full list of months and their agricultural activities may suggest the existence of a center for the study of the calendar. There too a row of upright monoliths indicates the existence in antiquity of a structure perhaps akin to that at Byblos; the remaining uprights, standing in a straight line, belie any circular arrangement.

The few known instances of monoliths arranged in a circle, somehow emulating the extraordinary circular arrangement at the Girsu, come to us from the Bible. Their rarity, however, points to a direct connection to Sumer in Gudea's time.

Knowledge of a circle of thirteen with an upright in the center emerges in the tale of Joseph, a great-grandson of Abraham, who kept annoying his eleven brothers by telling them of his dreams wherein they all bowed to him although he was the youngest. The dream that upset them most, leading them to get rid of him by selling him into slavery in Egypt, was the one in which, Joseph related, he saw "the Sun and the Moon and eleven stars bowing down to me," meaning his father and mother and eleven brothers.

Several centuries later, as the Israelites left Egypt for the Promised Land in Canaan, an actual stone circle—this time of twelve stones—was erected. In chapters 3 and 4 of the Book of Joshua the Bible describes the miraculous crossing of the Jordan River by the Israelites under the leadership of Joshua. As instructed by Yahweh, the heads of the twelve tribes erected twelve stones in the midst of the river; and as the priests carrying the Ark of the Covenant stepped into the waters and stood where the twelve stones were placed, the flow of the river's waters "was cut off" upstream and the dry river bed was exposed, enabling the Israelites to cross the Jordan on foot. As soon as the priests carrying the Ark stepped off the stones and carried the Ark across, "the waters of the Jordan returned to their place and flowed over its banks as they did before."

Then Yahweh ordered Joshua to take the twelve stones and erect them in a circle on the west side of the river, east of Jericho, as an everlasting commemoration of the miracle

performed by Yahweh. The place where the twelve stones
were erected was since then known as *Gilgal,* meaning
"Place of the Circle."

Not only the establishment of the twelve-stone circle as
a miraculous device is relevant here; so is the date of the
event. We first learn in Chapter 3 that the time was "harvest
time, when the waters of the Jordan overflow its banks."
Then Chapter 4 is more specific: it was in the first month
of the calendar, the month of the New Year; and it was on
the tenth of that month—the very day on which the inauguration ceremonies were culminated in Lagash—that "the
people left the Jordan and encamped at Gilgal, where Joshua
erected the twelve stones brought up from the Jordan
River."

These calendrical markers bear uncanny resemblance to
similar data concerning the time when Gudea had erected
the stone circles on the platform of the Girsu, after the
Eninnu itself was completed. We read in the Gudea inscriptions that the day Ninurta and his spouse entered their new
abode was the day when the harvest began in the land—
matching the "harvest time" in the tale of Gilgal. Astronomy and the calendar converge in both tales, and both concern circular structures.

The emergence of traditions of stone circles among the
descendants of Abraham can be traced, we believe, to Abraham himself and the identity of his father Terah. Dealing
with the subject in great detail in *The Wars of Gods and
Men,* we have concluded that Terah was an oracle priest of
royal descent, raised and trained in Nippur. Based on biblical data we have calculated that he was born in 2193 B.C.;
this means that Terah was an astronomer-priest in Nippur
when Enlil authorized his son, Ninurta, to proceed with the
building of the new Eninnu by Gudea.

Terah's son Abram (later renamed Abraham) was born,
by our calculations, in 2123 B.C. and was ten years old
when the family moved to Ur, where Terah was to serve
as a liaison. The family stayed there until 2096 B.C. when
it left Sumer for the Upper Euphrates region (a migration
that later led to Abraham's settlement in Canaan). Abraham

was by then well-versed in royal and priestly matters, including astronomy. Getting his education in the sacred precincts of Nippur and Ur just as the glories of the new Eninnu were talked about, he could not have missed learning of the wondrous stone circle of the Girsu; and this would explain the knowledge thereof by his descendants.

Where did the idea of a *circle* as a shape appropriate to astronomical observations—a shape that is the most outstanding feature of Stonehenge—come from? In our view, it came from the zodiac, the cycle of twelve constellations grouped around the Sun in the orbital plane (the Ecliptic) of the planets.

Earlier this century archaeologists uncovered in the Galilee, in northern Israel, the remains of synagogues dating to the decades and centuries immediately following the destruction of the Second Temple in Jerusalem by the Romans (in A.D. 70). To their surprise, a common feature of those synagogues was the decoration of their floors with intricate mosaic designs that included the signs of the zodiac. As this one from a place called Bet-Alpha shows (Fig. 89), the number—twelve—was the same as nowadays, the symbols were the same as now in use, and so were the names: written in a script no different from that of modern Hebrew, they begin (on the east) with *Taleh* for ram, Aries, flanked by *Shor* (bull) for Taurus and *Dagim* (fishes) for Pisces, and so on in the very same order that we continue to employ millennia later.

This zodiacal circle of what the Akkadians called *Manzallu* ("stations" of the Sun) was the source of the Hebrew term *Mazalot*, which came to denote "lucks." Therein lies the transition from the essential astronomical and calendrical nature of the zodiac to its astrological connotations—a transition that in time obscured the original significance of the zodiac and the role it played in the affairs of gods and men. Last but not least was its wondrous expression in the Eninnu that Gudea built.

The notion has prevailed, in spite of the facts, that the concept, names, and symbols of the zodiac were devised

Figure 89

by the Greeks, for the word is of Greek origin, meaning "animal circle." It is conceded that the inspiration for them may have come from Egypt, where the zodiac with its unaltered symbols, order, and names was certainly known (Fig. 90). In spite of the antiquity of some of the Egyptian depictions—including a magnificent one in the temple at Denderah, of which more later—the zodiac did not begin there. Studies such as the one by E.C. Krupp *(In Search of Ancient Astronomies)* have emphatically stated that "all available evidence indicates that the concept of the zodiac was not native to Egypt; instead, it is believed that the zodiac was imported to Egypt from Mesopotamia," at some unknown date. Greek savants, who had access to Egyptian art and traditions, had also attested in their writings that as far as astronomy was concerned, its knowledge came to them from the "Chaldeans," the astronomer-priests of Babylonia.

1. Aries. 2. Taurus. 3. Gemini.
4. Cancer. 5. Leo.
6. Virgo. 7. Libra. 8. Scorpio.
9. Sagittarius. 10. Capricorn.
11. Aquarius. 12. Pisces.

Figure 90

Archaeologists have found Babylonian astronomical tablets clearly marked off into twelve parts, each with its pertinent zodiacal symbol (Fig. 91). They may well represent the kind of sources that the Greek savants studied. Pictorially, however, the celestial symbols were carved on stones within a heavenly circle. Almost two thousands years before the circular zodiac of Bet-Alpha, Near Eastern rulers, especially in Babylon, invoked their gods on treaty documents; boundary stones *(Kudurru)* were emblazoned with the celestial symbols of these gods—planets and zodiacs—within the heavenly circle, embraced by an undulating serpent that represented the Milky Way (Fig. 92).

The zodiac, however, was begun, as far as Mankind is concerned, in Sumer. As we have undisputably shown in *The 12th Planet,* the Sumerians knew of, depicted (Fig. 93a) and named the zodiacal houses exactly as we still do six thousand years later:

Figure 91

Figure 92

GU.ANNA ("Heavenly Bull")—Taurus.
MASH.TAB.BA ("Twins")—Gemini.
DUB ("Pincers, Tongs")—the Crab (Cancer).
UR.GULA ("Lion")—Leo.
AB.SIN ("Whose Father was Sin")—the Maiden
  (Virgo).
ZI.BA.AN.NA ("Heavenly Fate")—the scales of
  Libra.
GIR.TAB ("The Clawer, the Cutter")—Scorpio.
PA.BIL ("Defender")—Archer (Sagittarius).
SUHUR.MASH ("Goat-fish")—Capricorn.
GU ("Lord of the Waters")—the Water Bearer
  (Aquarius).
SIM.MAH ("Fishes")—Pisces.
KU.MAL ("Field Dweller")—the Ram (Aries).

Overwhelming evidence demonstrates that the Sumerians were cognizant of the zodiacal ages—not only the names and images but the precessional cycle thereof—when the calendar was begun in Nippur, circa 3800 B.C., in the Age of Taurus. Willy Hartner, in his study titled "The Earliest History of the Constellations in the Near East" *(Journal of Near Eastern Studies)*, analyzed the Sumerian pictorial evidence and concluded that numerous depictions of a bull nudging a lion (Fig. 93b, from the fourth millenium B.C.) or a lion pushing bulls (Fig. 93c, from about 3000 B.C.) are representations of the zodiacal time when the spring equinox, at which time the calendrical new year began, was in the constellation Taurus and the summer solstice occurred in the sign of Leo.

Alfred Jeremias *(The Old Testament in the Light of the Ancient Near East)* found textual evidence that the Sumerian zodiacal-calendrical "point zero" stood precisely between the Bull and the Twins (Gemini), from which he concluded that the zodiacal division of the heavens—inexplicably to him—was devised even before the Sumerian civilization began, in the Age of Gemini. Even more puzzling to scholars has been a Sumerian astronomical tablet (VAT.7847 in the Berlin Vorderasiatisches Museum) that begins the list of zodiacal constellations with that of

Figures 93a, 93b, and 93c

Leo—taking one back to circa 11000 B.C., just about the time of the Deluge.

Devised by the Anunnaki as a link between Divine Time (the cycle based on the 3,600 years orbit of Nibiru) and Earthly Time (the Earth's orbital period), the Celestial Time (the time span of 2,160 years for the precessional shift from one zodiacal House to another) served to date major events in Earth's prehistory as archaeoastronomy could do in historical times. Thus, a depiction of the Anunnaki as astronauts and a spacecraft coursing between Mars (the six-pointed star) and Earth (identified by the seven dots and the accompanying crescent of the Moon) places the event, time-wise, in the Age of Pisces by including the zodiacal symbol of the two fishes in the depiction (Fig. 94). Written texts also included zodiacal dates; a text placing the Deluge in the Age of Leo is one example.

Even if we cannot be certain precisely when Mankind

Figure 94

was made aware of the zodiac, clearly it was long before Gudea's time. Hence it should not surprise us to discover that zodiacal depictions were indeed present in the new temple in Lagash; not, however, on the floor as in Bet-Alpha, and not as symbols carved on boundary stones. Rather, in a magnificent structure that can rightly be called *the first and most ancient planetarium!*

We read in Gudea's inscriptions that he emplaced "images of the constellations" in a "pure and guarded place, in an inner sanctuary." There, a specially designed "vault of heaven"—an imitation of the heavenly circle, a kind of ancient planetarium—was built as a dome resting on what is translated as "entablature" (a technical term meaning a base of a superstructure resting atop columns). In that "vault of heaven" Gudea "caused to dwell" the zodiacal images. We find clearly listed the "Heavenly Twins," the "Holy Capricorn," the "Hero" (Sagittarius), the Lion, the "Celestial Creatures" of the Bull and the Ram.

As Gudea had boasted, that "vault of heaven" studded with the zodiacal symbols must indeed have been a sight to behold. Millennia later, we can no longer step into that inner sanctum and share with Gudea the illusion of viewing the heavens with their shimmering constellations; but we could have gone to Denderah, in Upper Egypt, entered there the inner sanctum of its principal temple, and looked up to the ceiling. There we could have seen a painting of the starry heavens: the celestial circle, held up at the four cardinal points by the Sons of Horus and at the four points of solstitial sunrise and sunset by four maidens (Fig. 95). A circle depicting the thirty six "decans" (ten-day periods, three per month, of the Egyptian calendar) surrounds the central "vault of heaven" in which the twelve zodiacal constellations are depicted by the same symbols (bull, ram, lion, twins, etc.) and in the same order that we still use and that was begun in Sumer. The hieroglyphic name of the temple, *Ta ynt neterti*, meant "Place of the pillars of the goddess," suggesting that at Denderah too, as in the Girsu, stone uprights served for celestial observations, connected on the one hand to the zodiac and on the other hand to the calendar (as the thirty-six decans attest).

Figure 95

Scholars are unable to agree on the point in time represented by the Denderah zodiac. The depiction as now known was discovered when Napoleon visited Egypt, has since been removed to the Louvre Museum in Paris, and is believed to date to the period when Egypt came under Greco-Roman dominance. Scholars are, however, certain that it replicated a similar depiction in a much earlier temple, one that was dedicated to the goddess Hathor. Sir Norman Lockyer in *The Dawn of Astronomy* interpreted a Fourth Dynasty (2613–2494 B.C.) text as describing the celestial alignments in that earlier temple; this would date the Denderah ''vault of heaven'' to a time between the completion of Stonehenge I and the building of the Eninnu in Lagash by Gudea. If, as others hold, the skies shown in Denderah are dated by the image of the club topped by a falcon touching the foot of the Twins (Gemini), between the Bull (Taurus) on the right and the Crab (Cancer) on the left, it means that the Denderah depiction turned back the skies (as we do in modern planetariums when, say, at Christmas time the skies are shown as they were in the time of Jesus) to sometime between 6540 B.C. and 4380 B.C. According to the Egyptian chronology transmitted by the priests and recorded by Manetho, that was the time when demigods reigned over Egypt; such a dating of the Denderah skies (as distinct from when the temple itself was built) corroborates the findings, mentioned above, by Alfred Jeremias regarding the ''point zero'' of the Sumerian zodiacal calendar. Both Egyptian and Sumerian zodiacal datings thus confirm that the concept preceded the start of those civilizations, and that the ''gods,'' not men, were responsible for the depictions and their dating.

Since, as we have shown, the zodiac and its accompanying Celestial Time were devised by the Anunnaki soon after they first came to Earth, some of the zodiacal dates marking events depicted on cylinder seals do stand for zodiacal ages that preceded the emergence of Man's civilizations. The Age of Pisces, for example, that is indicated by the two fishes on Fig. 94, occurred no later than between 25980 B.C. and 23820 B.C. (or earlier if the event had taken place at prior ages of Pisces in the Great Cycle of 25,920 years).

Incredibly but not surprisingly, we find a suggestion that a "starry heaven" depicting the celestial circle with the constellations of the zodiac might have existed in the earliest times in a Sumerian text known to scholars as *A Hymn to Enlil the All-Beneficent*. Describing the innermost part of Enlil's Mission Control Center in Nippur, inside the E.KUR ziggurat, the text states that in a darkened chamber called Dirga there was installed "a heavenly zenith, as mysterious as the distant sea" in which "the starry emblems" were "carried to perfection."

The term DIR.GA connotes "dark, crownlike"; the text explains that the "starry emblems" installed therein enabled the determination of festivals, meaning a calendrical function. It all sounds like a forerunner of Gudea's planetarium; except that the one in the Ekur was hidden from human eyes, open to the Anunnaki alone.

Gudea's "vault of heaven," constructed as a planetarium, bears a greater similarity to the Dirga than to the depiction at Denderah, which was only a painting on the ceiling. Yet we cannot rule out the possibility that the inspiration for the one in the Girsu came from Egypt because of the numerous similarities to Egyptian ones that features in the Girsu bore. The list is far from being exhausted.

Some of the most impressive finds now adorning the Assyrian and Babylonian collections in the major museums are colossal stone animals with bodies of bulls or lions and heads of gods wearing horned caps (Fig. 96) that stood as guardians at temple entrances. We can safely assume that these "mythical creatures," as scholars call them, translated into stone sculptures the Bull-Lion motif that we illustrated earlier, thereby invoking for the temples the magic of an earlier Celestial Time and the gods associated with its past zodiacal ages.

Archaeologists believe that these sculptures were inspired by the sphinxes of Egypt, primarily the great Sphinx of Giza, with which the Assyrians and Babylonians were fa-

Figure 96

miliar as a result of both trade and warfare. But the Gudea inscriptions reveal that some fifteen hundred years before such zodiacal-cum-divine creatures were emplaced in Assyrian temples, Gudea had already positioned sphinxes at the Eninnu temple; the inscriptions specifically mention ''a lion that instilled terror'' and a ''wild ox, massively crouching like a lion.'' To the archaeologists' utter disbelief that sphinxes could have been known in ancient Sumer, a statue of Ninurta/Ningirsu himself, depicting him as a crouching sphinx (Fig. 97), was discovered among the ruins of the Girsu in Lagash.

Hints that all that should have been expected were given to Gudea—and thus to us—in the address by Ninurta to the baffled Gudea during the second night vision, in which Ninurta asserted his powers and reasserted his standing among the Anunnaki (''By fifty edicts my lordship is or-

Figure 97

dained''), pointed out his unusual familiarity with other parts of the world (''A lord whose eyes are lifted up afar'' as a result of his roamings in his Divine Black Bird), assured him of the cooperation of Magan and Melukhah (Egypt and Nubia), and promised him that the god called ''the Bright Serpent'' will, in person, come to assist in the construction of the new Eninnu: ''As a strong place it shall be built, like E.HUSH will my holy place be.''

This last statement is truly sensational in its implications.

''E'' as we already know meant a god's ''house,'' a temple; and in the case of the Eninnu—a stage-pyramid. HUSH (pronounced ''Chush'' with the ''ch'' as in the German *Loch*) meant in Sumerian ''of reddish hue, red-colored.'' So this is what Ninurta/Ningirsu stated: the new Eninnu will be like the ''Red-hued Divine House.'' The statement implies that the new Eninnu will emulate an existing structure known for its reddish hue...

Our search for such a structure can be facilitated by tracing back the pictograph for the sign *Hush*. What we find is truly astounding, for what it amounts to (Fig. 98a) is a line drawing of an *Egyptian pyramid* showing its shafts, internal passages, and subterranean chambers. More specifically, it appears to be drawn as a cross section of the Great Pyramid of Giza (Fig. 98b) and its trial scale model, the small pyramid of Giza (Fig. 98c)—and of the first successful phar-

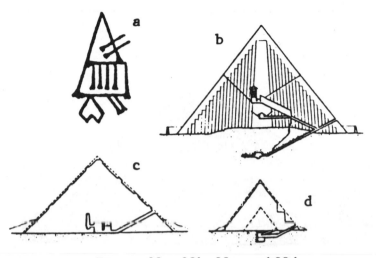

Figures 98a, 98b, 98c, and 98d

aonic pyramid (Fig. 98d) which, quite significantly—was called the *Red Pyramid,* of the very same hue that *Hush* had meant.

The Red Pyramid was certainly there to be emulated when the Eninnu was built in Lagash. It was one of three pyramids attributed to Sneferu, the first pharaoh of the IV dynasty, who reigned circa 2600 B.C. His architects first attempted to build for him a pyramid at Maidum, emulating the 52° slope of the Giza pyramids that were built millennia earlier by the Anunnaki; but the angle was too steep and the pyramid collapsed. The collapse led to a hurried change in the angle of a second pyramid at Dahshur to a flatter 43°, resulting in the pyramid nicknamed the Bent Pyramid. This led to the construction, also at Dahshur, of the third Sneferu pyramid. Considered the ''first classical pyramid'' of a pharaoh, its sides slope up at the safe angle of about 43½° (Fig. 99). It was built of local pink limestone and was therefore nicknamed the Red Pyramid. Protrusions on the sides were intended to hold in place a surfacing of white limestone; but that did not stay put for long, and today the pyramid is seen in its original reddish hue.

Having fought (and won) the Second Pyramid War in Egypt, Ninurta was not unfamiliar with its subsequent pyr-

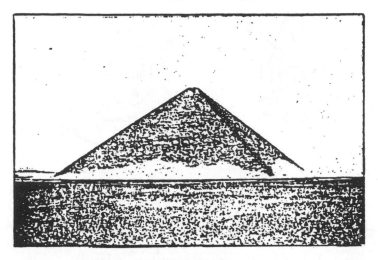

Figure 99

amids. Had he seen, as kingship came to Egypt, not only the Great Pyramid and its companions at Giza, but also the step-pyramid built by the Pharaoh Zoser at Sakkara, surrounded by its magnificent sacred precinct (see Fig. 78), built circa 2650 B.C.? Had he seen the final successful emulation by a pharaoh and his architects of the Great Pyramid—the Red Pyramid of Sneferu, built circa 2600 B.C.? And did he then tell the Divine Architect: that is what I would like to have built for me, a unique ziggurat combining elements of all three?

Else, how can one account for the compelling evidence linking the Eninnu, built between 2200 and 2100 B.C., with Egypt—and its gods?

And how else, except in this way, can one explain the similarities between Stonehenge in the British Isles and "Stonehenge on the Euphrates"?

For the explanation we have to turn our attention to the Divine Architect, the Keeper of the Secrets of the Pyramids, the god called by Gudea Ningishzidda; for he was none other than the Egyptian god *Tehuti* whom we call THOTH.

Thoth was called in the Pyramid Texts "He who reckons the heavens, the counter of the stars and the measurer of the Earth''; the inventor of arts and sciences, scribe of the gods, the "One who made calculations concerning the heavens, the stars and the Earth." As the "Reckoner of times and of seasons," he was depicted with a symbol combining the Sun's disk and the Moon's crescent upon his head, and—in words reminiscent of the biblical adoration of the Celestial Lord—the Egyptian inscriptions and legends said of Thoth that his knowledge and powers of calculating "measured out the heavens and planned the Earth." His hieroglyphic name *Tehuti* is usually explained as meaning "He who balances." Heinrich Brugsch *(Religion und Mythologie)* and E.A. Wallis Budge *(The Gods of the Egyptians)* interpreted that to mean that Thoth was the "god of the equilibrium" and considered depictions of him as "Master of the Balance" to indicate that he was associated with the equinoxes—the time when the day and the night were balanced. The Greeks identified Thoth with their god Hermes, whom they considered to have been the originator of astronomy and astrology, of the science of numbers and of geometry, of medicine and botany.

As we follow in the footsteps of Thoth, we shall come upon calendar tales that raise the curtain on the affairs of gods and men—and on enigmas such as Stonehenge.

8

CALENDAR TALES

The story of the calendar is one of ingenuity, of a sophisticated combination of astronomy and mathematics. It is also a tale of conflict, religious fervor, and struggles for supremacy.

The notion that the calendar was devised by and for farmers so that they would know when to sow and when to reap has been taken for granted too long; it fails both the test of logic and of fact. Farmers do not need a formal calendar to know the seasons, and primitive societies have managed to feed themselves for generations without a calendar. The historic fact is that the calendar was devised in order to predetermine the precise time of festivals honoring the gods. The calendar, in other words, was a religious device. The first names by which months were called in Sumer had the prefix EZEN. The word did not mean "month"; it meant "festival." The months were the times when the Festival of Enlil, or the Festival of Ninurta, or those of the other leading deities were to be observed.

That the calendar's purpose was to enable religious observances should not surprise one at all. We find an instance that still regulates our lives in the current common, but actually Christian, calendar. Its principal festival and the focal point that determines the rest of the annual calendar is Easter, the celebration of the resurrection, according to the New Testament, of Jesus on the third day after his crucifixion. Western Christians celebrate Easter on the first Sunday after the full moon that occurs on or right after the spring equinox. This created a problem for the early Christians in Rome, where the dominant calendrical element was the solar year of 365 days and the months were of irregular length and not exactly related to the Moon's phases. The determination of Easter Day therefore required a reliance on the Jewish calendar, because the Last Supper, from which the other crucial days of Eastertide are counted, was actually the *Seder* meal with which the Jewish celebration of Passover begins on the eve of the fourteenth day of the month Nissan, the time of the full Moon. As a result, during the first centuries of Christianity Easter was celebrated in accordance with the Jewish calendar. It was only when the Roman emperor Constantine, having adopted Christianity, convened a church council, the Council of Nicaea, in the year 325, that the continued dependence on the Jewish calendar was severed, and Christianity, until then deemed by the gentiles as merely another Jewish sect, was made into a separate religion.

In this change, as in its origin, the Christian calendar was thus an expression of religious beliefs and an instrument for determining the dates of worship. It was also so later on, when the Moslems burst out of Arabia to conquer by the sword lands and people east and west; the imposition of their purely lunar calendar was one of their first acts, for it had a profound religious connotation: it counted the passage of time from the *Hegira,* the migration of Islam's founder Mohammed from Mecca to Medina (in 622).

The history of the Roman-Christian calendar, interesting by itself, illustrates some of the problems inherent in the imperfect meshing of solar and lunar times and the resulting need, over the millennia, for calendar reforms and the ensuing notions of ever-renewing Ages.

The current Common Era Christian calendar was introduced by Pope Gregory XIII in 1582 and is therefore called the Gregorian Calendar. It constituted a reform of the previous Julian Calendar, so named after the Roman emperor Julius Caesar.

That noted Roman emperor, tired of the chaotic Roman calendar, invited in the first century B.C. the astronomer Sosigenes of Alexandria, Egypt, to suggest a reform of the calendar. Sosigenes's advice was to forget about lunar time-keeping and to adopt a solar calendar "as that of the Egyptians." The result was a year of 365 days plus a leap year of 366 days once in four years. But that still failed to account for the extra 11¼ minutes a year in excess of the quarter-day over and above the 365 days. That seemed too minute to bother with; but the result was that by 1582 the first day of spring, fixed by the Council of Nicaea to fall on March 21, was retarded by ten days to March 11th. Pope Gregory corrected the shortfall by simply decreeing on October 4, 1582, that the next day should be October 15. This reform established the currently used Gregorian calendar, whose other innovation was to decree that the year begin on January first.

The astronomer's suggestion that a calendar "as that of the Egyptians" be adopted in Rome was accepted, one must assume, without undue difficulty because by then Rome, and especially Julius Caesar, were quite familiar with Egypt, its religious customs, and hence with its calendar. The Egyptian calendar was at that time indeed a purely solar calendar of 365 days divided into twelve months of thirty days each. To these 360 days an end-of-year religious festival of five days was added, dedicated to the gods Osiris, Horus, Seth, Isis, and Nephthys.

The Egyptians were aware that the solar year is somewhat longer than 365 days—not just by the full day every four years, as Julius Caesar had allowed for, but by enough to shift the calendar back by one month every 120 years and by a full year every 1,460 years. The determining or sacred cycle of the Egyptian calendar was this 1,460-year period, for it coincided with the cycle of the heliacal rising of the star Sirius (Egyptian *Sept,* Greek *Sothis*) at the time of the Nile's annual flooding, which in turn takes place at about the summer solstice (in the northern hemisphere).

Edward Meyer (*Ägyptische Chronologie*) concluded that when this Egyptian calendar was introduced, such a convergence of the heliacal rising of Sirius and of the Nile's inundation had occurred on July 19th. Based on that Kurt Sethe (*Urgeschichte und älteste Religion der Ägypter*) calculated that this could have happened in either 4240 B.C.

or 2780 B.C. by observing the skies at either Heliopolis or Memphis.

By now researchers of the ancient Egyptian calendar agree that the solar calendar of 360 + 5 days was not the first prehistoric calendar of that land. This "civil" or secular calendar was introduced only after the start of dynastic rule in Egypt, i.e., after 3100 B.C.; according to Richard A. Parker (*The Calendars of the Ancient Egyptians*) it took place circa 2800 B.C. "probably for administrative and fiscal purposes." This civil calendar supplanted, or perhaps supplemented at first, the "sacred" calendar of old. In the words of the *Encyclopaedia Britannica,* "the ancient Egyptians originally employed a calendar based upon the Moon." According to R.A. Parker (*Ancient Egyptian Astronomy*) that earlier calendar was, "like that of all ancient peoples," a calendar of twelve *lunar* months plus a thirteenth intercalary month that kept the seasons in place.

That earlier calendar was also, in the opinion of Lockyer, equinoctial and linked indeed to the earliest temple at Heliopolis, whose orientation was equinoctial. In all that, as in the association of months with religious festivals, the earliest Egyptian calendar was akin to that of the Sumerians.

The conclusion that the Egyptian calendar had its roots in predynastic times, before civilization appeared in Egypt, can only mean that it was not the Egyptians themselves who invented their calendar. It is a conclusion that matches that regarding the zodiac in Egypt, and regarding both the zodiac and the calendar in Sumer: they were all the artful inventions of the "gods."

In Egypt, religion and worship of the gods began in Heliopolis, close by the Giza pyramids; its original Egyptian name was *Annu* (as the name of the ruler of Nibiru) and it is called *On* in the Bible: when Joseph was made viceroy over all of Egypt (Genesis chapter 41), the Pharaoh "gave him Assenath, the daughter of Potiphera, the [high] priest of On, for a wife." Its oldest shrine was dedicated to *Ptah* ("The Developer") who, according to Egyptian tradition, raised Egypt from under the waters of the Great Flood and made it habitable by extensive drainage and earthworks. Divine reign over Egypt was then transferred by Ptah to his son *Ra* ("The Bright One") who was also called *Tem* ("The Pure One"); and in a special shrine, also at Heliopolis, the Boat of Heaven of Ra, the conical *Ben-Ben,* could be seen by pilgrims once a year.

Ra was the head of the first divine dynasty according to the Egyptian priest Manetho (his hieroglyphic name meant "Gift of Thoth"), who compiled in the third century B.C. Egypt's dynastic lists. The reign of Ra and his successors, the gods Shu, Geb, Osiris, Seth, and Horus, lasted more than three millennia. It was followed by a second divine dynasty that was begun by Thoth, another son of Ptah; it lasted half as long as the first divine dynasty. Thereafter a dynasty of demigods, thirty of them, reigned over Egypt for 3,650 years. Altogether, according to Manetho, the divine reigns of Ptah, the Ra dynasty, the Thoth dynasty, and the dynasty of the demigods lasted 17,520 years. Karl R. Lepsius (*Königsbuch der alten Ägypter*) noted that this time span represented exactly twelve Sothic cycles of 1,460 years each, thereby corroborating the prehistoric origin of calendrical-astronomical knowledge in Egypt.

Based on substantial evidence, we have concluded in *The Wars of Gods and Men* and other volumes of *The Earth Chronicles* that Ptah was none other than Enki and that Ra was Marduk of the Mesopotamian pantheon. It was to Enki and his descendants that the African lands were granted when Earth was divided among the Anunnaki after the Deluge, leaving the E.DIN (the biblical land of Eden) and the Mesopotamian sphere of influence in the hands of Enlil and his descendants. Thoth, a brother of Ra/Marduk, was the god the Sumerians called Ningishzidda.

Much of the history and violent conflicts that followed the Earth's division stemmed from the refusal of Ra/Marduk to acquiesce in the division. He was convinced that his father was unjustly deprived of lordship of Earth (what the epithet-name EN.KI, "Lord Earth," connoted); and that therefore he, not Enlil's Foremost Son Ninurta, should rule supreme on Earth from Babylon, the Mesopotamian city whose name meant "Gateway of the Gods." Obsessed by this ambition, Ra/Marduk caused not only conflicts with the Enlilites, but also aroused the animosity of some of his own brothers by involving them in these bitter conflicts as well as by leaving Egypt and then returning to reclaim the lordship over it.

In the course of these comings and goings and ups and downs in Ra/Marduk's struggles, he caused the death of a younger brother called Dumuzi, let his brother Thoth reign and then forced him into exile, and made his brother Nergal change sides in a War of the Gods that resulted in a nuclear holocaust. It was in particular the on-again, off-again relationship with Thoth, we believe, that is essential to the Calendar Tales.

The Egyptians, it will be recalled, had not one but two calendars. The first, with roots in prehistoric times, was "based upon the Moon." The later one, introduced several centuries after the start of pharaonic rule, was based on the 365 days of the solar year. Contrary to the notion that the latter "civil calendar" was an administrative innovation of a pharaoh, we suggest that it too, like the earlier one, was an artful creation of the gods; except that while the first one was the handiwork of Thoth, the second one was the craftwork of Ra.

One aspect of the civil calendar considered specific and original to it was the division of the thirty-day months into "decans," ten-day periods each heralded by the heliacal rising of a certain star. Each star (depicted as a celestial god sailing the skies, Fig. 100) was deemed to give notice of the last hour of the night; and at the end of ten days, a new decan-star would be observed.

It is our suggestion that the introduction of this decan-based calendar was a deliberate act by Ra in a developing conflict with his brother Thoth.

Both were sons of Enki, the great scientist of the Anunnaki, and one can safely assume that much of their

Figure 100

knowledge had been acquired from their father. This is certain in the case of Ra/Marduk, for a Mesopotamian text has been found that clearly states so. It is a text whose beginning records a complaint by Marduk to his father that he lacks certain healing knowledge. Enki's response is rendered thus:

> *My son, what is it you do not know?*
> *What more could I give to you?*
> *Marduk, what is it that you do not know?*
> *What could I give you in addition?*
> *Whatever I know, you know!*

Was there, perhaps, some jealousy between the two brothers on this score? The knowledge of mathematics, of astronomy, of orienting sacred structures was shared by both; witness to Marduk's attainments in these sciences was the magnificent ziggurat of Babylon (see Fig. 33) which, according to the *Enuma elish,* Marduk himself had designed. But, as the above-quoted text relates, when it came to medicine and healing, his knowledge fell short of his brother's: he could not revive the dead, while Thoth could. We learn of the latter's powers from both Meso-potamian and Egyptian sources. His Sumerian depictions show him with the emblem of the entwined serpents (Fig. 101a), the emblem originally of his father Enki as the god

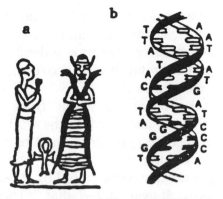

Figures 101a and 101b

who could engage in genetic engineering—the emblem, we have suggested, of the double helix of DNA (Fig. 101b). His Sumerian name, NIN.GISH.ZID.DA, which meant "Lord of the Artifact of Life," bespoke recognition of his capacity to restore life by reviving the dead. "Lord healer, Lord who seizes the hand, Lord of the Artifact of Life" a Sumerian liturgical text called him. He was prominently featured in magical healing and exorcism texts; a *Maqlu* ("Burnt Offerings") series of incantations and magical formulas devoted a whole tablet, the seventh, to him. In one incantation, devoted to drowned mariners ("the seafaring folk who are utterly at rest"), the priest invokes the formulas of "Siris and Ningishzidda, the miracle workers, the spellbinders."

Siris is the name of a goddess otherwise unknown in the Sumerian pantheon, and the possibility that it is a Mesopotamian rendition of the star's name Sirius comes to our mind because in the Egyptian pantheon Sirius was the star associated with the goddess Isis. In Egyptian legendary tales, Thoth was the one who had helped Isis, the wife of Osiris, to extract from the dismembered Osiris the semen with which Isis was impregnated to conceive and bear Horus. This was not all. In an Egyptian inscription on an artifact known as the Metternich Stela, the goddess Isis describes how Thoth brought her son Horus back from the dead after Horus was stung by a poisonous scorpion. Responding to her cries, Thoth came down from the skies, "and he was provided with magical powers, and possessed the great power which made the word become indeed." And he performed magic, and by nighttime it drove the poison away and Horus was returned to life.

The Egyptians held that the whole *Book of the Dead*, verses from which were inscribed on the walls of pharaonic tombs so that the deceased pharaoh could be translated into an Afterlife, was a composition of Thoth, "written with his own fingers." In a shorter work called by the Egyptians the *Book of Breathings*, it was stated that "Thoth, the most mighty god, the lord of Khemennu, cometh to thee; he writeth for thee the Book of Breathings with his own fingers, so that thy *Ka* shall breathe for ever and ever and thy form endowed with life on Earth."

We know from Sumerian sources that this knowledge, so essential in pharaonic beliefs—knowledge to revive the dead—was first possessed by Enki. In a long text dealing with Inanna/Ishtar's journey to the Lower World (southern Africa), the domain of her sister who was married to another son of Enki, the uninvited goddess was put to death. Responding to appeals, Enki fashioned medications and supervised the treatment of the corpse with sound and radiation pulses, and "Inanna arose."

Evidently, the secret was not divulged to Marduk; and when he complained, his father gave him an evasive answer. That alone would have been enough to make the ambitious and power-hungry Marduk jealous of Thoth. The feeling of being offended, perhaps even threatened, was probably greater. First, because it was Thoth, and not Marduk/Ra, who had helped Isis retrieve the dismembered Osiris (Ra's grandson) and save his semen, and then revived the poisoned Horus (a great-grandson of Ra). And second, because all that led—as the Sumerian text makes clearer—to an affinity between Thoth and the star Sirius, the controller of the Egyptian calendar and the harbinger of the life-giving inundation of the Nile.

Were these the only reasons for the jealousy, or did Ra/Marduk have more compelling reasons to see in Thoth a rival, a threat to his supremacy? According to Manetho, the long reign of the first divine dynasty begun by Ra ended abruptly after only a short reign of three hundred years by Horus, after the conflict that we have called the First Pyramid War. Then, instead of another descendant of Ra, it was Thoth who was given lordship over Egypt and his dynasty continued (according to Manetho) for 1,570 years. His reign, an era of peace and progress, coincided with the New Stone (Neolithic) Age in the Near East—the first phase of the granting of civilization by the Anunnaki to Mankind.

Why was it Thoth, of all the other sons of Ptah/Enki, who was chosen to replace the dynasty of Ra in Egypt? A clue might be suggested in a study titled *Religion of the Ancient Egyptians* by W. Osborn, Jr., in which it is stated as follows regarding Thoth: "Though he stood in mythology in a secondary rank of deities, yet he always remained a direct emanation from, and part of, Ptah—the *firstborn* of the primeval deity" (emphasis is ours). With the complex rules of succession of the Anunnaki, where a son born to a half sister became the legal heir ahead of a firstborn son (if mothered not by a half sister)—a cause of the endless friction and rivalry between Enki (the first-born of Anu) and Enlil (born to a half sister of Anu)— could it be that the circumstances of Thoth's birth somehow posed a challenge to Ra/Marduk's claims for supremacy?

It is known that initially the dominating "company of the gods" or divine dynasty was that of Heliopolis; later on it was superseded by the divine triad of Memphis (when Memphis became the capital of a unified Egypt). But in between there was an interim *Paut* or "divine company" of gods headed by Thoth. The "cult center" of the latter was Hermopolis ("City of Hermes" in Greek) whose Egyptian name, *Khemennu*, meant "eight." One of the epithets of Thoth was "Lord of Eight," which according to Heinrich Brugsch (*Religion und Mythologie der alten Aegypter*) referred to eight celestial orientations, including the four cardinal points. It could also refer to Thoth's ability to ascertain and mark out the eight standstill points of the Moon—the celestial body with which Thoth was associated.

Marduk, a "Sun god," on the other hand, was associated with the number ten. In the numerical hierarchy of the Anunnaki, in which Anu's rank was the highest, sixty, that of Enlil fifty and of Enki forty (and so on down), the rank of Marduk was ten; and that could have been the origin of the decans. Indeed, the Babylonian version of the Epic of Creation attributes to Marduk the devising of a calendar of twelve months each divided into three "celestial astrals":

> *He determined the year,*
> *designating the zones:*
> *For each of the twelve months*
> *he set up three celestial astrals,*
> *[thus] defining the days of the year.*

This division of the skies into thirty-six portions as a means of "defining the days of the year" is as clear a reference as possible to the calendar—a calendar with thirty-six "decans." And here, in *Enuma elish*, the division is attributed to Marduk, alias Ra.

The Epic of Creation, undoubtedly of Sumerian origin, is known nowadays mostly from its Babylonian rendition (the seven tablets of the *Enuma elish*). It is a rendition, all scholars agree, that was intended to glorify the Babylonian national god Marduk. Hence, the name "Marduk" was inserted where in the Sumerian original text the invader from outer space, the planet Nibiru, was described as the Celestial Lord; and where, describing deeds on Earth, the Supreme God was named Enlil, the Babylonian version also named Marduk. Thereby, Marduk was made supreme both in heaven and on Earth.

Without further discovery of intact or even fragmented tablets inscribed with the original Sumerian text of the Epic of Creation, it is impossible to say whether the thirty-six decans were a true innovation by Marduk or were just borrowed by him from Sumer. A basic tenet of Sumerian astronomy was the division of the celestial sphere enveloping the Earth into three "ways"—the Way of Anu as a central celestial band, the Way of Enlil of the northern skies, and the Way of Ea (i.e., Enki) in the southern skies. It has been thought that the three ways represented the equatorial band in the center and the bands demarcated by the two tropics, north and south; we have, however, shown in *The 12th Planet* that the Way of Anu, straddling the equator, extended 30° northward and southward of the equator, resulting in a width of 60°; and that the Way of Enlil and the Way of Ea similarly extended for 60° each, so that the three covered the complete celestial sweep of 180° from north to south.

If this tripartite division of the skies were to be applied to the calendrical division of the year into twelve months, the result would be thirty-six segments. Such a division—resulting in decans—was indeed made, in Babylon.

In 1900, addressing the Royal Astronomical Society in London, the orientalist T.G. Pinches presented a reconstruction of a Mesopotamian astrolabe (literally: "Taker of stars"). It was a circular disk divided like a pie into twelve segments and three concentric rings, resulting in a division of the skies into thirty-six portions (Fig. 102). The round symbols next to the inscribed names indicated that the reference was to celestial bodies; the names (here transliterated) are those of constellations of the zodiac, stars, and planets—thirty-six in all. That this division was linked to the calendar is made clear by the inscribing of the months' names, one in each of the twelve segments at the segment's top (the marking I to XII, starting with the first month Nisannu of the Babylonian calendar, is by Pinches).

While this Babylonian planisphere does not answer the question of the origin of the relevant verses in *Enuma elish*, it does establish that what was supposed to have been a unique and original Egyptian innovation in fact had a counterpart (if not a predecessor) in Babylon—the place claimed by Marduk for his supremacy.

Even more certain is the fact that the thirty-six decans

Figure 102

do not feature in the first Egyptian calendar. The earlier one was linked to the Moon, the later one to the Sun. In Egyptian theology, Thoth was a Moon God, Ra was a Sun God. Extending this to the two calendars, it follows that the first and older Egyptian calendar was formulated by Thoth and the second, later one, by Ra/Marduk.

The fact is that when the time came, circa 3100 B.C., to extend the Sumerian level of civilization (human Kingship) to the Egyptians, Ra/Marduk—having been frustrated in his efforts to establish supremacy in Babylon—returned to Egypt and expelled Thoth.

It was then, we believe, that Ra/Marduk—not for administrative convenience but in a deliberate step to eradicate the vestiges of Thoth's predominance—reformed the calendar. A passage in the *Book of the Dead* relates that Thoth was "disturbed by what hath happened to the divine children" who have "done battle, upheld strife, created fiends, caused trouble." As a consequence of this Thoth "was provoked to anger when they [his adversaries] bring the years to confusion, throng in and push to disturb the months." All that evil, the text declares, "in all they have done unto thee, they have worked iniquity in secret."

This may well indicate that the strife that led to the substitution of Thoth's calendar by Ra/Marduk's calendar in Egypt took place when the calendar (for reasons explained earlier) needed to be put back on track. R.A. Parker, we have noted above, believes that this change occurred circa 2800 B.C. Adolf Erman (*Aegypten und Aegyptisches Leben im Altertum*) was more specific. The opportunity, he wrote, was the return of Sirius to its original position, after the 1,460-year cycle, on July 19, 2776 B.C.

It should be noted that that date, circa 2800 B.C., is the official date adopted by the British authorities for Stonehenge I.

The introduction by Ra/Marduk of a calendar divided into, or based upon, ten-day periods may have also been prompted by a desire to draw a clear distinction, for his followers in Egypt as well as in Mesopotamia, between himself and the one who was "seven"—the head of the Enlilites, Enlil himself. Indeed, such a distinction may have underlain the oscillations between lunar and solar calendars; for the calendars, as we have shown and ancient records attested, were devised by the Anunnaki "gods" to delineate for their followers the cycles of worship; and the struggle for supremacy meant, in the final analysis, who was to be worshiped.

Scholars have long debated, but have yet to verify, the origin of the week, the slice of the year measured in lengths of seven days. We have shown in earlier books of *The Earth Chronicles* that seven was the number that represented our planet, the Earth. Earth was called in Sumerian texts "the seventh," and was depicted in representations of celestial bodies by the symbol of the seven dots (as in Fig. 94) because journeying into the center of our Solar System from their outermost planet, the Anunnaki would first encounter Pluto, pass by Neptune and Uranus (second and third), and continue past Saturn and Jupiter (fourth and fifth). They would count Mars as the sixth (and therefore it was depicted as a six-pointed star) and Earth would be the seventh. Such a journey and such a count are in fact depicted on a planisphere discovered in the ruins of the royal library of Nineveh, where one of its eight segments (Fig. 103) shows the flight path from Nibiru and states (here in English translation) "deity Enlil went by the planets." The planets, represented by dots, are seven in number. For the Sumerians, it was Enlil, and no other, who was "Lord of Seven." Mesopotamian as well as biblical names, of persons (e.g., *Bath-sheba*, "Daughter of Seven") or of places (e.g., *Beer-Sheba*, "the well of Seven") honored the god by this epithet.

The importance or sanctity of the number seven, transferred to the calendrical unit of seven days as one week, permeates the Bible and other ancient scriptures. Abraham set apart seven ewe lambs when he negotiated with Abimelech; Jacob served Laban seven years to be able to marry one of his daughters, and bowed seven times as he approached his jealous brother Esau. The High Priest was

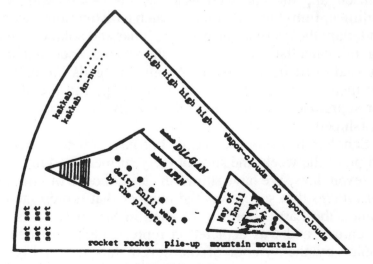

Figure 103

required to perform various rites seven times, Jericho was to be circled seven times so that its walls should tumble down; and calendrically, the seventh day had to be strictly observed as the Sabbath and the important festival of Pentecost had to take place after the count of seven weeks from Passover.

Though no one can say who "invented" the seven-day week, it is obviously associated in the Bible with the earliest times—indeed, when Time itself began: witness the seven days of Creation with which the book of Genesis begins. The concept of a seven-day delineated period of counted time, a Time of Man, is found in the biblical as well as the earlier Mesopotamian Deluge tale, thereby attesting to its antiquity. In the Mesopotamian texts, the hero of the flood is given seven days' advance warning by Enki, who "opened the water clock and filled it" to make sure his faithful follower would not miss the deadline. In those versions the Deluge is said to have begun with a storm that "swept the country for seven days and seven nights." In the biblical version the Deluge also began after a seven-day advance warning to Noah.

The biblical tale of the flood and its duration reveals a far-reaching understanding of the calendar in very early times. Significantly, it shows familiarity with the unit of seven days and of a division of the year into fifty-two weeks of seven days each. Moreover, it suggests an understanding of the complexities of a lunar-solar calendar.

According to Genesis, the Deluge began "in the second month, on the seventeenth day of the month" and ended the following year "in the second month, on the twenty-seventh day of the month." But what on the face of it would appear to be a period of 365 days plus ten, is not so. The biblical tale breaks down the Deluge into 150 days of the avalanche of water, 150 days during which the water receded, and another forty days until Noah deemed it safe to open the Ark. Then, in two seven-day intervals, he sent out a raven and a dove to survey the landscape; only when the dove no longer came back did Noah know it was safe to step out.

According to this breakdown, it all added up to 354 days (150 + 150 + 40 + 7 + 7). But that is not a solar year; that is precisely a lunar year of twelve months averaging 29.5 days each (29.5 × 12 = 354) represented by a calendar—as the Jewish one still is—alternating between months of 29 and 30 days.

But 354 days is not a full year in solar terms. Recognizing this, the narrator or editor of Genesis resorted to intercalation, by stating that the Deluge, which began on the seventeenth day of the second month, ended (a year later) on the twenty-seventh of the second month. Scholars are divided in regard to the number of days thus added to the lunar 354. Some (e.g., S. Gandz, *Studies in Hebrew Mathematics and Astronomy*) consider the addition to have been eleven days—the correct intercalary addition that would have expanded the lunar 354 days to the full 365 days of the solar year. Others, among them the author of the ancient *Book of Jubilees*, consider the number of days added to be just ten, increasing the year in question to only 364 days. The significance is, of course, that it implies a calendar divided into fifty-two weeks of seven days each (52 × 7 = 364).

That this was not just a result of adding 354 + 10 as the number of days, but a deliberate division of the year into fifty-two weeks of seven days each, is made clear in the text of the *Book of Jubilees*. It states (chapter 6) that Noah was given, when the Deluge ended, "heavenly tablets" ordaining that

> *All the days of the commandment*
> *will be two and fifty weeks of days*
> *which will make the year complete.*
> *Thus it is engraven and ordained*
> *on the heavenly tablets;*
> *there shall be no neglecting for a single*
> *year or from year to year.*
> *And command thou the children of Israel*
> *that they observe the years according to*
> *this reckoning:*

three hundred and sixty-four days;
these shall constitute a complete year.

The insistence on a year of fifty-two weeks of seven days, adding up to a calendrical year of 364 days, was not a result of ignorance regarding the true length of 365 full days in a solar year. The awareness of this true length is made clear in the Bible by the age (*"five* and sixty and three hundred years"*)* of Enoch until he was lofted by the Lord. In the nonbiblical *Book of Enoch* the "overplus of the Sun," the five epagomenal days that had to be added to the 360 days (12 × 30) of other calendars, to complete the 365, are specifically mentioned. Yet the *Book of Enoch*, in chapters describing the motions of the Sun and the Moon, the twelve zodiacal "portals," the equinoxes and the solstices, states unequivocally that the calendar year shall be "a year exact as to its days: three hundred and sixty-four." This is repeated in a statement that "the complete year, with perfect justice" was of 364 days—fifty-two weeks of seven days each.

The *Book of Enoch*, especially in its version known as Enoch II, is believed to show elements of scientific knowledge centered at the time in Alexandria, Egypt. How much of that can be traced back to the teachings of Thoth cannot be stated with any certainty; but biblical as well as Egyptian tales suggest a role for seven and fifty-two times seven beginning in much earlier times.

Well known is the biblical tale of Joseph's rise to governorship over Egypt after he had successfully interpreted the pharaoh's dreams of, first, seven fatfleshed cows that were devoured by seven leanfleshed cows, and then of seven full ears of corn swallowed up by seven dried-out ears of corn. Few are aware, however, that the tale—"legend" or "myth" to some—had strong Egyptian roots as well as an earlier counterpart in Egyptian lore. Among the former was the Egyptian forerunner of the Greek Sibylline oracle goddesses; they were called the Seven Hathors, Hathor having been the goddess of the Sinai peninsula who was depicted as a cow. In other words, the Seven

Hathors symbolized seven cows who could predict the future.

The earlier counterpart of the tale of seven lean years that followed seven years of plenty is a hieroglyphic text (Fig. 104) that E.A.W. Budge (*Legends of the Gods*) titled "A legend of the god Khnemu and of a seven year famine." Khnemu was another name for Ptah/Enki in his role as fashioner of Mankind. The Egyptians believed that after he had turned over lordship over Egypt to his son Ra, he retired to the island of Abu (known as Elephantine since Greek times because of its shape), where he formed twin caverns—two connected reservoirs—whose locks or sluices could be manipulated to regulate the flow of the Nile's waters. (The modern Aswan High Dam is similarly located above the Nile's first cataract).

According to this text, the Pharaoh Zoser (builder of

Figure 104

the step-pyramid at Saqqara) received a royal dispatch from the governor of the people of the south that grievous suffering had come upon the people "because the Nile hath not come forth to the proper height for *seven years*." As a result, "grain is very scarce, vegetables are lacking altogether, every kind of thing which men eat for their food hath ceased, and every man now plundereth his neighbor."

Hoping that the spread of famine and chaos could be avoided by a direct appeal to the god, the king traveled south to the island of Abu. The god, he was told, dwells there "in an edifice of wood with portals formed of reeds," keeping with him "the cord and the tablet" that enable him to "open the double door of the sluices of the Nile." Khnemu, responding to the king's pleadings, promised "to raise the level of the Nile, give water, make the crops grow."

Since the annual rising of the Nile was linked to the heliacal rising of the star Sirius, one must wonder whether the tale's celestial or astronomical aspects recall not only the actual shortage of water (which occurs cyclically even nowadays) but also to the shift (discussed above) in the appearance of Sirius under a rigid calendar. That the whole tale had calendrical connotations is suggested by the statement in the text that the abode of Khnemu at Abu was astronomically oriented: "The god's house hath an opening to the southeast, and the Sun standeth immediately opposite thereto every day." This can only mean a facility for observing the Sun in the course of moving to and from the winter solstice.

This brief review of the use and significance of the number seven in the affairs of gods and men suffices to show its celestial origin (the seven planets from Pluto to Earth) and its calendrical importance (the seven-day week, a year of fifty-two such weeks). But in the rivalry among the Anunnaki, all that assumed another significance: the determination of who was the God of Seven (*Eli-Sheva* in Hebrew, from which Elizabeth comes) and thus the titular Ruler of Earth.

And that, we believe, is what alarmed Ra/Marduk on his return to Egypt after his failed coup in Babylon: the spreading veneration of Seven, still Enlil's epithet, through the introduction of the seven-day week into Egypt.

In these circumstances the veneration of the Seven Hathors, as an example, must have been anathema to Ra/Marduk. Not only their number, seven, which implied veneration of Enlil; but their association with Hathor, an important deity in the Egyptian pantheon but one for whom Ra/Marduk had no particular liking.

Hathor, we have shown in earlier books of *The Earth Chronicles,* was the Egyptian name for Ninharsag of the Sumerian pantheon—a half sister of both Enki and Enlil and the object of both brothers' sexual attention. Since the official spouses of both (Ninki of Enki, Ninlil of Enlil) were not their half sisters, it was important for them to beget a son by Ninharsag; such a son, under the succession rules of the Anunnaki, would be the undisputed Legal Heir to the throne on Earth. In spite of repeated attempts by Enki, all Ninharsag bore him were daughters; but Enlil was more successful, and his Foremost Son was conceived in a union with Ninharsag. This entitled Ninurta (Ningirsu, the "Lord of Girsu" to Gudea) to inherit his father's rank of fifty— at the same time depriving Enki's firstborn, Marduk, of rulership over the Earth.

There were other manifestations of the spread of the worship of seven and its calendrical importance. The tale of the seven-year drought takes place at the time of Zoser, builder of the Saqqara pyramid. Archaeologists have discovered in the area of Saqqara a circular "altar-top" of alabaster whose shape (Fig. 105) suggests that it was intended to serve as a sacred lamp to be lighted over a seven-day period. Another find is that of a stone "wheel" (some think it was the base of an omphalos, an oracular "navel stone") that is clearly divided into four segments of seven markers each (Fig. 106), suggesting that it was really a stone calendar—a lunar calendar, no doubt— incorporating the seven-day week concept and (with the

Figure 105

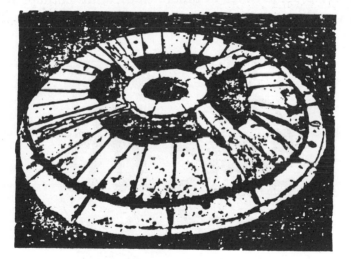

Figure 106

aid of the four dividers) enabling a lunar monthly count ranging from twenty-eight to thirty-two days.

Calendars made of stone had existed in antiquity, as evidence Stonehenge in Britain and the Aztec calendar in Mexico. That this one was found in Egypt should be the least wonder, for it is our belief that the genius behind all of those geographically spread stone calendars was one and the same god: Thoth. What may be surprising is this calendar's embracing the cycle of seven days; but that too, as another Egyptian "legend" shows, should not have been unexpected.

What archaeologists identify as games or game boards have been found almost everywhere in the ancient Near East, as witness these few illustrations of finds from Mesopotamia, Canaan, and Egypt (Fig. 107). The two players moved pegs from one hole to another in accordance with the throw of dice. Archaeologists see in that no more than games with which to while away the time; but the usual number of holes, fifty-eight, is clearly an allocation of twenty-nine to each player—and twenty-nine is the number of full days in a lunar month. There were also obvious subdivisions of the holes into smaller groups, and grooves connected some holes to others (indicating perhaps that the player could jump-advance there). We notice, for example,

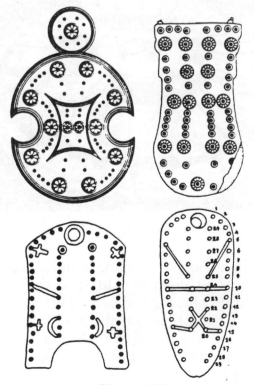

Figure 107

that hole 15 was connected to hole 22 and hole 10 to 24, which suggests a "jump" of one week of seven days and of a fortnight of fourteen days.

Nowadays we employ ditties ("Thirty days hath September") and games to teach the modern calendar to children; why exclude the possibility that it was so also in antiquity?

That these were calendar games and that at least one of them, the favorite of Thoth, was designed to teach the division of the year into fifty-two weeks, is evident from an ancient Egyptian tale known as "The Adventures of Satni-Khamois with the Mummies."

It is a tale of magic, mystery, and adventure, an ancient thriller that combines the magical number fifty-two with Thoth and the secrets of the calendar. The tale is written on a papyrus (cataloged as Cairo-30646) that was discovered in a tomb in Thebes, dating to the third century B.C. Fragments of other papyruses with the same tale have also been found, indicating that it was part of the established or canonical literature of ancient Egypt that recorded the tales of gods and men.

The hero of this tale was Satni, a son of the pharaoh, "well instructed in all things." He was wont to wander in the necropolis of Memphis, studying the sacred writings on temple walls and researching ancient "books of magic." In time he himself became "a magician who had no equal in the land of Egypt." One day a mysterious old man told him of a tomb "where there is deposited the book that the god Thoth had written with his own hand," and in which the mysteries of the Earth and the secrets of heaven were revealed. That secret knowledge included divine information concerning "the risings of the Sun and the appearances of the Moon and the motions of the celestial gods [the planets] that are in the cycle [orbit] of the Sun"; in other words—the secrets of astronomy and the calendar.

The tomb in question was that of Ne-nofer-khe-ptah, the son of a former king. When Satni asked to be shown the location of this tomb, the old man warned him that although

Nenoferkheptah was buried and mummified, he was not dead and could strike down anyone who dared take away the Book of Thoth that was lodged at his feet. Undaunted, Satni searched for the subterranean tomb, and when he reached the right spot he "recited a formula over it and a gap opened in the ground and Satni went down to the place where the book was."

Inside the tomb Satni saw the mummies of Nenoferkheptah, of his sister-wife, and of their son. The book was indeed at Nenoferkheptah's feet, and it "gave off a light as if the sun shone there." As Satni stepped toward it, the wife's mummy spoke up, warning him to advance no further. She then told Satni of her own husband's adventures when he had attempted to obtain the book, for Thoth had hidden it in a secret place, inside a golden box that was inside a silver box that was inside a series of other boxes within boxes, the outermost ones being of bronze and iron. When her husband, Nenoferkheptah, ignored the warnings and the dangers and grasped the book, Thoth condemned him and his wife and their son to suspended animation: although alive, they were buried; and although mummified, they could see, hear, and speak. She warned Satni that if he touched the book, his fate would be the same or worse.

The warnings and the fate of the earlier king did not deter Satni. Having come so far, he was determined to get the book. As he took another step toward the book, the mummy of Nenoferkheptah spoke up. There was a way to possess the book without incurring the wrath of Thoth, he said: to play and win the Game of Fifty-Two, "the magical number of Thoth."

Challenging fate, Satni agreed. He lost the first game and found himself partly sunk into the floor of the tomb. He lost the next game, and the next, sinking down more and more into the ground. How he managed to escape with the book, the calamities that befell him as a result, and how he in the end returned the book to its hiding place, makes fascinating reading but is unessential to our immediate subject: the fact that the astronomical and calendrical "secrets of Thoth'' included the Game of Fifty-Two—the division of the year into fifty-two seven-day portions, resulting in the enigmatic year of only 364 days of the books of Jubilees and Enoch.

It is a magical number that vaults us across the oceans, to the Americas, returns us to the enigma of Stonehenge, and parts the curtains on the events leading to, and resulting from, the first New Age recorded by Mankind.

9

WHERE THE SUN ALSO RISES

No view epitomizes Stonehenge more than the sight of the Sun's rays shining through the still-standing megaliths of the Sarsen Circle at sunrise on summer's longest day, when the Sun in its northern migration seems to hesitate, stop, and begin to return. As fate would have it, only four of those great stone pillars remain upright and connected at the top by the curving lintels, forming three elongated windows through which we, as though we were Stonehenge's long-gone giant builders, can also view—and determine— the beginning of a new annual cycle (Fig. 108).

And as fate would have it, somewhere on the other side of the world, another set of three windows in a massive structure of cyclopean stones—built, local lore relates, by giants—also offers a breathtaking view of the Sun appearing through white and misty clouds to direct its rays in a precise alignment. That other place of the Three Windows, where the Sun also rises on a crucial calendrical day, is in South America, in Peru (Fig. 109).

Is the similarity just a visual fluke, a mere coincidence? We think not.

Nowadays the place is called Machu Picchu, so named after the sharp peak that rises ten thousand feet at a bend of the Urubamba River on which the ancient city is situated. So well hidden in the jungle and among the endless peaks of the Andes, it eluded the Spanish *Conquistadors* and re-mained a "lost city of the Incas" until discovered in 1911 by Hiram Bingham. It is now known that it was built long before the Incas, and that its olden name was *Tampu-Tocco*, "Haven of the Three Windows." The place, and its unique three windows, are featured in local lore regarding the origins of the Andean civilization when the gods, led by

Figure 108

Figure 109

the great creator Viracocha, placed the four Ayar brothers and their four sister-wives in Tampu-Tocco. Three brothers emerged through the three windows to settle and civilize the Andean lands; one of them founded the Ancient Empire that preceded that of the Incas by thousands of years.

The three windows formed part of a massive wall constructed of cyclopean granite stones that—as at Stonehenge—were not native to the site, but hauled from a great distance across towering mountains and steep valleys. The colossal stones, carefully smoothed and rounded on their surfaces, were cut into numerous sides and angles as though they were soft putty. Each stone's sides and angles fitted the sides and angles of all its adjoining stones; all these polygonal stones thus locked into one another like pieces of a jigsaw puzzle, tightly fitting without any mortar or cement and withstanding the not infrequent earthquakes in the area and other ravages of man and nature.

The Temple of the Three Windows, as Bingham named it, has only three walls: the one with the windows facing in an easterly direction, and two sidewalls as protecting wings. The western side is completely open, providing room for a stone pillar, about seven feet high; supported by two horizontally placed, carefully shaped stones, one on each side, the pillar precisely faces the central window. Because of a niche cut into the pillar's top, Bingham surmised that it might have held a beam supporting a thatched roofing; but that would have been a unique feature in Machu Picchu, and we believe that the pillar here served the same purpose as the Heel Stone (at first) at Stonehenge or the Altar Stone (later on there), i.e., as the Seventh Pillar of Gudea to provide the line of sight. Ingeniously, the availability of three windows made possible three lines of sight—to sunrise on midsummer day, equinox day, and midwinter day (Fig. 110).

The structure of the three windows with the facing pillar made up the eastern part of what Bingham named, and scholars still call, the Sacred Plaza. Its other principal structure, also three-sided, has its longest wall on the Plaza's northern end and is without a wall on its southern face. It

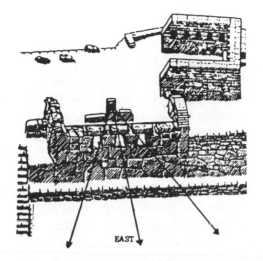

EAST

Figure 110

too is made of cyclopean blocks of imported granite also held together by their polygonal shapes. The central north wall has been so constructed as to create seven false windows—trapezoidal cutouts that imitate the three windows but do not in fact cut through the stone wall. A massive rectangular stone monolith, measuring fourteen by five by three feet, lies on the structure's floor below these false windows. Though the purpose of this structure has not been determined, it is still referred to as the Principal Temple, as Bingham named this structure.

Since the five-foot height of the prostrate stone did not let it serve as a seat, Bingham speculated that it might have served as an offering table, "a species of an altar; possibly offerings of food were placed on it, or it may have been intended to receive mummies of the honored dead, which could here be brought out and worshipped on days of festival." Though such customs are purely imaginary, the suggestion that the structure could have been related to festival days—i.e. to the calendar—is intriguing. The false seven windows have six markedly protruding stone pegs above them, so that some kind of counting involving seven and six—as at the Girsu in Lagash—cannot be ruled out. The two sidewalls contain five false windows each, so that each sidewall—one on the east, one on the west—provided a count of twelve together with the central (northern) wall. This too implies a calendrical function.

A smaller enclosure that belongs to the same Megalithic Age was built as an adjunct to the Principal Temple, behind its northwest corner. It can best be described as a roofless room with a stone bench; Bingham assumed that it was the priest's abode, but there is nothing there to indicate its purpose. What is obvious, though, is that it was built with the greatest care of the same polygonal granite boulders, shaped and polished to perfection. Indeed, it is there that the stone with the most sides and angles—thirty-two!—is found; how and by whom this amazing megalith was carved and emplaced is a mystery that confounds the visitor.

Right behind this enclosure there begins a stairway, made of rectangularly shaped but undressed field stones that serve as steps. It winds its way upward, leading from the Sacred Plaza up a hill which overlooks the whole city. The top of the hill was flattened to enable the construction of an enclosure. It was also built of beautifully shaped and smoothed stones, but not of a megalithic size and not outstandingly polygonal; rather, the higher entrance wall that creates a gateway to the hilltop and the surrounding lower walls are built of ashlars—rectangularly shaped stones that, as bricks, form masonry walls. This method of construction is neither of a kind with the colossi of the Megalithic Age nor of the obviously inferior structures of field stones, mortared together in their irregular shapes, of which most other structures at Machu Picchu are built. The latter undoubtedly belong to the Inca period; and the ashlar-built structures, as the one on the hilltop, belong to an earlier era which, in *The Lost Realms,* we have identified as the era of the Ancient Empire.

The ashlar-built structure atop the hill was clearly intended only as a decorative-protective enclosure for the main feature of the hilltop. There, in the center, where the hilltop was flattened to form a platform, an outcropping of the native stone was left sticking out, then shaped and carved magnificently to create a polygonal base from which a short stone column projects upward. That the stone-on-a-base served astronomical-calendrical purposes is evident from its name: *Inti-huatana,* which in the local tongue meant "That which binds the Sun." As the Incas and their descendants explained, it was a stone instrument for observing and determining the solstices, to make sure that the Sun be bound and not keep moving away for good without being pulled back to return (Fig. 111).

Nearly a quarter of a century passed between the discovery of Machu Picchu and the first serious study of its astronomical connotations. It was only in the 1930s that Rolf Müller, a professor of astronomy at the University of Potsdam in Germany, began a series of investigations at several important sites in Peru and Bolivia. Fortunately, he applied to his findings the principles of archaeoastronomy that were first expounded by Lockyer; and so, besides the interesting conclusions regarding the astronomical aspects of Machu Picchu, Cuzco, and Tiahuanacu (on the southern shores of Lake Titicaca), Müller was able to pinpoint their time of construction.

Müller concluded (*Die Intiwatana (Sonnenwarten) im Alten Peru* and other writings) that the short pillar atop the base, and the base itself, were cut and shaped to enable

Figure 111

precise astronomical observations at this particular geographical location and elevation. The pillar (Fig. 112a)
served as a gnomon and the base as a recorder of the shadow.
However, the base itself was so shaped and oriented that
observations along its grooves could pinpoint sunrise or
sunset on crucial days (Fig. 112b). Müller concluded that
those preintended days were sunset (*Su*) on the day of the
winter solstice (June 21 in the southern hemisphere) and
sunrise (*Sa*) on the day of the summer solstice (there, December 23). He furthermore determined that the angles of
the rectangular base were such that if one were to observe
the horizon along a diagonal sightline connecting protrusions 3 and 1, one would have observed sunset precisely
on the equinox days at the time the Intihuatana was carved.

That, he concluded based on the Earth's greater tilt at the
time, was just over four thousand years ago—sometime

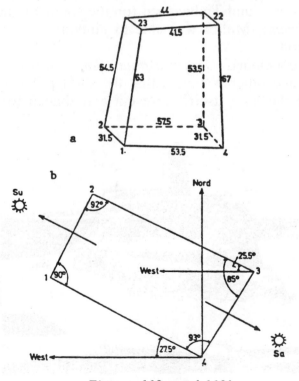

Figures 112a and 112b

between 2100 B.C. and 2300 B.C. This makes the Intihuatana at Machu Picchu contemporaneous with, if not somewhat older than, the Eninnu in Lagash and Stonehenge II. More remarkable perhaps is the rectangular layout for the astronomical function of the Intihuatana's base, for it imitates the exceptional rectangular layout of the four Station Stones of Stonehenge I (though, apparently, without its lunar purposes).

The legend of the Ayar brothers relates that the three brothers from whom the Andean kingdoms stemmed—a kind of South American version of the biblical Ham, Shem, and Japhet—got rid of the fourth brother by imprisoning him in a cave inside a great rock, where he was turned into a stone. Such a cave inside a cleft great rock, with a white vertical stem or short pillar inside, indeed exists at Machu Picchu. Above it one of the most remarkable structures in the whole of South America still stands. Built of the same kind of ashlars as on the platform of the Intihuatana, and thus clearly contemporaneous with it, is an enclosure which on two sides forms perfect walls at a right angle to each other, and on the other two sides curves to form a perfect semicircle (Fig. 113a). It is known the *Torreon* (the Tower).

The enclosure, which is reached by seven stone steps, encompasses, as at the Intihuatana, the protruding peak of the great rock on which it was constructed. As with the Intihuatana, the outcropping here was also carved and given a purposeful shape; except that here no stem was made to act as a gnomon. Instead, the astronomical sightlines that run along grooves and polygonal surfaces of the "sacred rock" lead to two windows in the semicircular wall. Müller, and other astronomers after him (e.g., D.S. Dearborn and R.E. White, *Archaeoastronomy at Machu Picchu*), concluded that the sightlines were oriented to sunrises on the days of the winter and summer solstices—more than four thousand years ago (Fig. 113b).

The two windows were similar in their trapezoid shape (wider at the bottom, narrower at the top) to the legendary Three Windows in the Sacred Plaza and thus emulated, in shape and purpose, the ones from the Megalithic Age. The

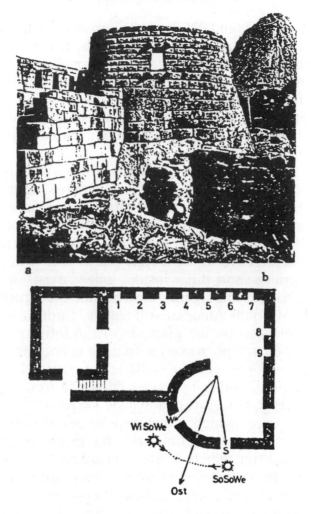

Figures 113a and 113b

similarity continued in that the structure of perfect ashlars, where the semicircle ended and the northerly straight wall began, had a third window—if one can so call the aperture. It is larger than the other two; its sill is not straight, but is shaped as an inverted stairway; and its top is formed not by a straight lintel stone but by a wedgelike slit, like an inverted V (Fig. 114).

Because the view through this opening (from inside the Torreon outward) is obstructed by fieldstone buildings from

Figure 114

Inca times, the astronomers who had studied the Torreon attached no astronomical significance to this Third Window. Bingham pointed out that the wall at this window showed clear evidence of fire, and he surmised that it was evidence of the burning of sacrifices on certain festival dates. Our own studies show that when the Incan buildings were not yet there, i.e., at the time of the Ancient Empire, a sightline from the Sacred Rock through the slit in this window to the Intihuatana atop the hill to the northwest would probably have indicated winter solstice sunset when the Torreon was built.

The structure atop the cleft rock also emulated those in the Sacred Plaza in other features. In addition to the three apertures, there were nine false trapezoidal windows in the straight parts of the enclosing walls (see Fig. 113). Spaced between these false windows there protrude from the walls stone pegs or "bobbins" as Bingham called them (Fig. 115). The longer wall, which has seven false windows, has six such pegs—duplicating the arrangement in the longer wall of the Principal Temple.

The number of the windows—actual plus false—twelve, undoubtedly denotes calendrical functions, such as the count of twelve months in a year. The number of false windows (seven) and pegs (six) in the longer wall, as in that of the Principal Temple, may indicate a calendrical need to engage in intercalation—a periodic adjustment of the lunar cycle to the solar cycle by adding a thirteenth month every few years. Combined with the alignments and apertures for observing and determining the solstices and the equinoxes, the false windows with their pegs lead to the conclusion that at Machu Picchu someone had created a complex solar-lunar stone computer to serve as a calendar.

The Torreon, contemporaneous with the Eninnu and with Stonehenge II, is in one respect more remarkable than the rectangular format at the Intihuatana, because it presents the extremely rare *circular* shape of a stone structure— extremely rare, that is, in South America, but with an obvious kinship to the stone circles of Lagash and Stonehenge.

Figure 115

According to legends and data compiled by the Spaniard Fernando Montesinos at the beginning of the seventeenth century, the Inca Empire was not the first kingdom with a capital at Cuzco in Peru. Researchers now know that the legendary Incas, whom the Spaniards encountered and subjugated, came to power at Cuzco only in A.D. 1021. Long before them one of the Ayar brothers, Manco Capac, founded the city when a golden rod given him by the god Viracocha sank into the ground to indicate the right location. It happened, by the calculations of Montesinos, circa 2400 B.C.—almost 3,500 years before the Incas. That Ancient Empire lasted nearly 2,500 years until a succession of plagues, earthquakes, and other calamities caused the people to leave Cuzco. The king, accompanied by a handful of chosen people, retreated to the hideout of Tampu-Tocco; there, the interregnum lasted about a thousand years, until a young man of noble birth was chosen to lead the people back to Cuzco and establish a New Kingdom—that of the Inca dynasty.

When the Spanish conquerors arrived in Cuzco, the Inca capital, in 1533, they were astounded to discover a metropolis with some 100,000 dwelling houses surrounding a royal-religious center of magnificent palaces, temples, plazas, gardens, marketplaces, parade grounds. They were puzzled to hear that the city was divided into twelve wards, arranged in an oval, whose boundaries ran along sightlines anchored to observation towers built on peaks encircling the city (Fig. 116). And they were awed by the sight of the city's and empire's holiest temple—not because it was superbly built, but because it was literally covered with gold. True to its name *Cori-cancha,* meaning Golden Enclosure, the temple's walls were covered with plates of gold; inside there were wondrous artifacts and sculptures of birds and animals made of gold, silver, and precious stones; and in the temple's main courtyard there was a garden whose corn and other growths were all artificial, made of gold and silver. The initial scouting party of Spaniards alone removed seven hundred of those gold plates (as well as many of the other precious artifacts).

Figure 116

Chroniclers who had seen the Coricancha before it was vandalized, demolished by the Catholic priests, and built over into a church, reported that the enclosed compound included a main temple, dedicated to the god Viracocha; and shrines or chapels for the worship of the Moon, Venus, a mysterious star called *Coyllor,* the Rainbow, and the god of Thunder and Lightning. The Spaniards nevertheless called the temple Temple of the Sun, believing that the Sun was the supreme deity wcrshiped by the Incas.

It is assumed that the idea came to the Spaniards from the fact that in the Holy of Holies of the Coricancha—a semicircular chamber—there hung on the wall above the great altar an "image of the Sun." It was a great golden disk which the Spaniards assumed to represent the Sun. In reality, it had served in earlier times to reflect a beam of light as the Sun's rays penetrated the dark chamber once a year—at the moment of sunrise on the day of the winter solstice.

Significantly, the arrangement was akin to that in the Great Temple of Amon at Karnak, in Egypt. Significantly, the Holy of Holies was in the same extremely rare form of a semicircle, as the Torreon in Machu-Picchu. Significantly, the earliest part of the temple, including the Holy of Holies, was built of the same perfect ashlars as the Torreon and the walls enclosing the Intihuatana—the hallmark of the Ancient Empire era. And, not surprisingly, careful studies and measurements by Müller showed that the orientation designed to permit the beam of sunlight to travel through the corridor and bounce off the ''image of the Sun'' was conceived when the Earth's obliquity was 24° (Fig. 117), which chronologically means, he wrote, more than four thousand years earlier. This matches the timetable related by Montesinos, according to which the Ancient Empire began circa 2500–2400 B.C. and the assertion that the temple in Cuzco was built soon thereafter.

As astoundingly early as the structures of the Ancient Empire were, they were clearly not the earliest ones, for according to the Ayar legends the megalithic Three Windows had already existed when the founder of the Ancient Empire, Manco Capac, and his brothers set out from Tampu-Tocco to establish kingships in the Andean lands.

A Megalithic Age with its colossal structures had ob-

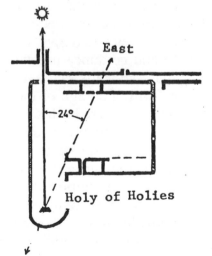

Figure 117

viously preceded the Ancient Empire—structures distinguished not only by their immense size, but also by the amazing polygonality of their stone blocks coupled with the smoothly shaped and somewhat rounded faces of these megaliths. But as mind-boggling as that age's structures at Machu Picchu are, they are neither the largest nor the most enigmatic ones. The prize for that should undoubtedly go to the ruins at Sacsahuaman, the promontory that overlooks Cuzco.

Shaped like a triangle with its base toward the mountain chain of which the promontory is an edge, its two sides formed by deep gorges, its apex forms a peak that rises steeply some eight hundred feet above the city that lies at its bottom. The promontory can be divided into three parts. The widest part, forming the triangle's base, is dominated by huge rock outcroppings that someone—"giants" according to local lore—had cut and shaped, with incredible ease and at angles that could not possibly have been formed with crude hand tools, to form giant steps or platforms or inverted stairs, additionally perforating the rocks with twisting channels, tunnels, grooves, and niches. The promontory's middle section is formed by an area hundreds of feet wide and long that has been flattened out to form a huge level area. This flattened-out area is clearly separated from the triangular and more elevated apex of the promontory by a most remarkable and certainly unique stone structure. It consists of three massive walls that extend in a zigzag parallel to each other from one edge of the promontory to the other (Fig. 118). The walls are built so as to rise one behind the other, to a combined height of about sixty feet. They are constructed of the colossal stone blocks and in the polygonal fashion that is the earmark of the Megalithic Age; those in the forefront, that support the earthfills that form the raised terraces for the second and third tiers, are the most massive. Its smallest boulders weigh between ten and twenty tons; most are fifteen feet high and are ten to fourteen feet in width and thickness. Many are much larger; one boulder in the front row is twenty-seven feet high and weighs over three hundred tons (Fig. 119). As with the other mega-

Figure 118

Figure 119

liths at Machu Picchu, the ones at Sacsahuaman too were brought over from a great distance, were given their smooth and beveled faces and polygonal shapes, and remain holding fast together without mortar.

By whom, when, and why were these structures above ground, and the tunnels, channels, conduits, bored holes, and other odd shapes carved into the living rocks, made and fashioned? Local lore attributed it to "the giants." The Spaniards, as the chronicler Garcilaso de la Vega wrote, believed that they were "erected not by men but by demons." Squier wrote that the zigzagging walls represented "without doubt the grandest specimens of the style called Cyclopean extant in America," but offered no explanation or theory.

Recent excavations have uncovered behind the large rock outcroppings that separate the flat middle area from the rocky area leading to the northwest, where most of the tunnels and channels have been formed, one of the most unusual structural shapes in South America: a perfect circle. Carefully shaped stones have been laid out so as to form the rim of a sunken area, perfectly circular. In *The Lost Realms* we enumerate the reasons for our conclusion that it served as a reservoir where ores—gold ores, to be specific—were processed as in a giant pan.

This, however, was not the only circular structure on the promontory. Assuming that the three tiers of colossal walls were ramparts of a fortress, the Spaniards took it for granted that structural remains in the highest and narrowest part of the promontory, behind and above the walls, belonged to an Inca fortification. Prompted by local legends that a child once fell into a hole there and later emerged eight hundred feet down in Cuzco proper, local archaeologists engaged in limited excavations. They discovered that the area behind and above the three walls was honeycombed with subterranean tunnels and chambers. More important, they uncovered there the foundations of a series of connected square and rectangular buildings (Fig. 120a); in their midst there were the remains of a perfectly circular structure. The natives refer to the structure as the *Muyocmarca*, "The Circular Building"; the archeologists call it the *Torreon*—the Tower—the same descriptive name given to the semicircular structure at Machu Picchu, and assumed that it was a defensive tower, part of the Sacsahuaman "fortress."

Archaeoastronomers, however, see in the structure clear evidence of an astronomical function. R.T. Zuidema (*Inca Observations of the Solar and Lunar Passages*, and other studies) noted that the alignment of the straight walls adjoining the circular structure was such that the north and south points of zenith and nadir could have been determined there. The walls that form the square enclosure within which

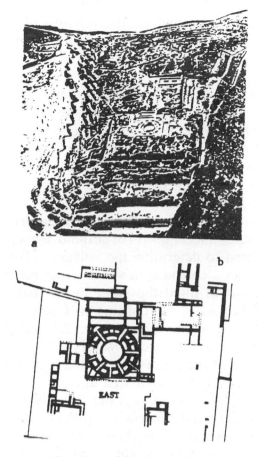

Figures 120a and 120b

the circular structure was emplaced are indeed aligned with the cardinal points (Fig. 120b); but they form only a frame for the circular structure, which consisted of three concentric walls connected by spokes of masonry that divide the outer two circular walls into sections. One such opening—an aperture if the higher courses forming the tower followed the ground plan—does point due south and thus could have served to determine sunset on nadir day. But the four other openings are clearly oriented to the northeast, southeast, southwest, and northwest—the unmistakable points of sunset and sunrise on the winter and summer solstice days (in the southern hemisphere).

If these are, at it appears, the remains of a full-fledged astronomical observatory, it was in all probability *the earliest round observatory in South America,* perhaps in all of the Americas.

The alignment of this round observatory to the solstices puts it in the same category as the one at Stonehenge and orientationally as that of Egyptian temples. The evidence suggests, however, that after the Megalithic Age and in the era of the Ancient Empire begun under the aegis of Viracocha, both the equinoxes and the lunar cycle played the key roles in the Andean calendar.

The chronicler Garcilaso de la Vega, describing the towerlike structures (see Fig. 116) around Cuzco, stated that they were used to determine the solstices. But he also described another "calendar in stone" that has not survived and that brings to mind the stone circle that stood on the platform in Lagash . . . According to Garcilaso, the pillars erected in Cuzco served to determine the equinoxes, not the solstices. These are his exact words: "In order to denote the precise day of the equinoctial, pillars of the finest marble had been erected in the open area in front of the Coricancha which, when the Sun came near the time, the priests watched daily to observe what shadow the pillars cast; and to make it more exact they fixed on them a gnomon like the pin of a dial. And so, soon as the sun at its rising came to cast a direct shadow by it and at the sun's height, by midday, cast no shadow, they concluded that the sun had entered the equinoctial."

According to the authoritative study *The Andean Calendar* by L.E. Valcarcel, such a fixing and veneration of the equinoxes was carried into Inca times although they switched from an earlier equinoctial calendar to a solstitial one. His study brought out the fact that Inca month names accorded special significance to months corresponding to our March and September, the equinoctial months. "The Incas believed," he wrote, "that on the two days of the equinoxes Father Sun came down to live among men."

The need to adjust the solar calendar over a period of millennia because of the phenomenon of precession and, perhaps, also due to the wavering between a solstitial and an equinoctial New Year, led to repeated reforms of the calendar even in the days of the Ancient Empire. According to Montesinos, the 5th, 22nd, 33rd, 39th, and 50th monarchs of the Ancient Empire "renewed the computation of time that had fallen into confusion." That such calendar reforms had to do with wavering between solstices and equinoxes is confirmed by the statement that the monarch Manco Capac IV "ordered that the year begin at the spring equinox," a feat possible because he was an *Amauta,* a "knower of astronomy." But evidently in doing so he only reinstated a calendar that had once been in use, in earlier times; for, according to Montesinos, the fortieth monarch who had reigned a thousand years before Manco Capac IV, "established an academy for the study of astronomy and determined the equinoxes. He was knowing in astronomy and found the equinoxes, which the Indians called *Illa-Ri.*"

As if all that was not enough to require constant reforms, other evidence also indicates the employment of, or at least familiarity with, the lunar calendar. In his studies of Andean archaeoastronomy Rolf Müller reported that at a site called Pampa de Anta, some ten miles west of Sacsahuaman, the sheer rock has been carved into a series of steps that form a semicircle or crescent. Since there is nothing to view there except the promontory at Sacsahuaman to the east, Müller concluded that the place served to make astronomical observations along a sightline anchored on the Sacsahuaman promontory—but, apparently, linked to appearances of the Moon. The native name for the edifice, *Quillarumi,* "Moon Stone," suggests such a purpose.

Shackled by the notions that the Incas worshiped the Sun, modern scholars found it at first difficult to concede that Inca observations could have also included the Moon. In fact the early Spanish chroniclers stated repeatedly that the Incas had an elaborate and precise calendar incorporating both solar and lunar aspects. The chronicler Felipe Guaman Poma de Avila stated that the Incas "knew the cycles of the Sun and the Moon . . . and the month of the year and the four winds of the world." The assertion that the Incas observed both solar and lunar cycles is confirmed by the fact that next to the shrine to the Sun in the Coricancha there was a shrine to the Moon. In the Holy of Holies the central symbol was an ellipse flanked by the Sun on the left and the Moon on the right; it was only the ruler Huascar, one of the two half brothers who were fighting over the throne when the Spaniards arrived, who replaced the oval with a golden disk representing the Sun.

These are Mesopotamian calendrical features; finding them in the remote Andes has baffled the scholars. Even more perplexing has been the certainty that the Incas were familiar with the zodiac—a wholly arbitrary device for dividing the orbital circle around the Sun into twelve parts— a Sumerian "first" by all accounts.

E.G. Squier, in his report on Cuzco and the meaning of its name ("Navel of the Earth"), noted that the city was divided into twelve wards arranged around the nucleus or "navel" in an elliptical shape (Fig. 121), which is the true orbital circuit. Sir Clemens Markham (*Cuzco and Lima: the Incas of Peru*) quoted the chronicler Garcilaso de la Vega's information that the twelve wards represented the twelve zodiacal constellations. Stansbury Hagar (*Cuzco, the Celestial City*) noted that, according to Inca lore, Cuzco was laid out in conformity with a sacred or divine plan to emulate the heavens, and concluded that the first ward, named "Terrace of Kneeling," represented the constellation Aries. He showed that—as in Mesopotamia—the Incas also associated each of the twelve zodiac "houses" with a parallel month in the calendar. These zodiacal months bore names that had an uncanny resemblance to their Near Eastern names that originated in Sumer. Thus, the month of the autumn equinox, which equaled the month of the spring equinox and the constellation of the Bull (Taurus) when the calendar began in Sumer, was called *Tupa Taruca*, "Pasturing Stag." The constellation of the Maiden (Virgo), as another example, was called *Sara Mama*, "Mother Maize." To grasp fully the extent of such similarities, one should recall that in Mesopotamia this constellation (see Fig. 91) was depicted as a maiden holding a stalk of grain—wheat or barley

Figure 121

in Mesopotamia, replaced by *maiz* (corn) in the Andes. Hagar's conclusion that the zodiacal layout in Cuzco associated the first ward with Aries rather than with Taurus as in Sumer, suggests that the city's plan was devised after the Age of Taurus had ended (due to precession) at about 2150 B.C. According to Montesinos, it was the fifth ruler of the Ancient Empire who completed the Coricancha and introduced a new calendar some time after 1900 B.C. That *Capac* (ruler) was given the epithet *Pachacuti* (Reformer), and one can safely conclude that the reform of the calendar in his time was required by the zodiacal shift from Taurus to Aries—another confirmation of familiarity with the zodiac and its calendrical aspects even in pre-Inca times in the Andes.

There were other aspects—complex aspects—of the ancient Near Eastern calendars in the calendar that the Incas had retained from the days of the Ancient Empire. The requirement (still in force in the Jewish and Christian calendars) that the spring festival (Passover, Easter) be held when the Sun is in the relevant zodiac house *and* on or immediately after the first full Moon of that month, forced the ancient priest-astronomers to intercalate the solar and lunar cycles. The studies by R.T. Zuidema and others concluded that not only did such intercalation take place in the Andes, but that the lunar cycle was additionally linked to two other phenomena: it had to be the first full Moon after the June solstice, and it was to coincide with the first heliacal rising of a certain star. This double correlation is intriguing, for it brings to mind the Egyptian linking of the beginning of their calendrical cycle both to the solar date (rising of the Nile) and the heliacal rising of a star (Sirius).

Some twenty miles northeast of Cuzco, at a place called Pisac, there are remains of a structure, probably from early Inca times, that appear to have been an attempt to emulate and combine some of the sacred structures at Machu Picchu: a building one of whose sides was semicircular, with a crude *Intihuatana* in its midst. At a place not far from Sacsahuaman called Kenko, a large semicircle of well-shaped ashlars fronts on a large stone monolith that could have had the shape of an animal (the features are too damaged to be discerned); whether or not this edifice had astronomical-calendrical functions is unknown. These sites, added to those of Machu Picchu, Sacsahuaman, and Cuzco, illustrate the fact that in what has been called the Sacred Valley— *and only there*—religion, the calendar, and astronomy led to the construction of *circular or semicircular* observatories; nowhere else in South America do we find such structures.

Who was it who, *at about the same time,* applied the same set of astronomical principles and adopted a circular shape for celestial observations in early Britain, at Lagash in Sumer, and in South America's Ancient Empire?

All legends, supported by geographical evidence and archaeological finds, point to the southern shores of Lake Titicaca as the place of the South American Beginning—

not only of human civilization, but of the gods themselves. It was there, according to the legends, that the repopulation of the Andean lands began after the Deluge; that the gods, headed by Viracocha, had their abode; that the couples destined to begin the Ancient Empire were given knowledge, route instructions, and the Golden Wand with which to locate the site of the Navel of the Earth—of establishing *Cuzco*.

Insofar as human beginnings in the Andes are concerned, the tales connected them to two distinct islands off the southern shore of Lake Titicaca. They were called the Island of the Sun and the Island of the Moon, the two luminaries having been considered as the two principal helpers of Viracocha; the calendrical symbolism inherent in these tales has been noted by many scholars. The abode of Viracocha was, however, in a City of the Gods on the mainland, at the lake's southern shore. The place, called Tiahuanacu, was settled by the gods (according to local lore) in times immemorial; it was, the legends related, a place of colossal structures that only giants could erect.

The chronicler Pedro Cieza de León, who traveled throughout what is now Peru and Bolivia in the years immediately following the Spanish conquest, reported that without doubt, of all the antiquities in the Andean lands, the ruins at Tiahuanacu were "the most ancient place of any." Among the edifices that amazed him was an artificial hill "on a great stone foundation" that measured more than 900 feet by 400 feet at its base and rose some 120 feet. Nearby he saw gigantic stone blocks fallen to the ground, among them "many doorways with their jambs, lintels, and thresholds all in one stone" which in turn were part of even larger stone blocks, "some of them thirty feet broad, fifteen or more long, and six in thickness." He wondered whether "human force can have sufficed to move them to the place where we see them, being so large." But not only the immense size of the stone blocks puzzled him; so did their "grandeur and magnificence." "For myself," he wrote, "I fail to understand with what instruments or tools it could have been done, for it is certain that before these great stones could be brought to perfection and left as we see them, the tools must have been much better than those now used by the Indians.'' He had no doubt that ''two stone idols, of the human shape and figure, the features very skillfully carved . . . that seem like small giants'' were responsible for the wondrous structures.

Over the centuries most of the smaller stone blocks have been carted away to be used in La Paz, the Bolivian capital, in railroads leading to it, and in rural areas all around. But even so, travelers continued to report the incredible monumental remains; by the end of the nineteenth century the reports assumed a more scientific accuracy as a result of the visits and researches by Ephraim George Squier (*Peru: Incidents of Travel and Exploration in the Land of the Incas*) and A. Stübel and Max Uhle (*Die Ruinenstaette von Tiahuanaco im Hochland des Alten Peru*). They were followed earlier this century by the most renowned and tenacious researcher of Tiahuanacu, Arthur Posnansky (*Tiahuanacu— The Cradle of American Man*). Their work and more recent excavations and studies, reviewed at length in *The Lost Realms,* have led us to conclude that Tiahuanacu was the tin capital of the ancient world, that its extensive aboveground and underground structures were metallurgical facilities, that the huge one-piece multiwalled stone blocks were part of port facilities at the ancient lakeshore, and that Tiahuanacu was founded not by Man but by the Anunnaki ''gods'' in their search for gold long before Man was taught the uses of tin.

Where a narrow and rare plain fanned out from the southern shore of Lake Titicaca, the site of the once magnificent Tiahuanacu and its port (nowadays called Puma-Punku), only three principal monuments to its past dominate the landscape. The one at the southeastern part of the ruins is the hill called *Akapana,* an artificial hill (as Cieza de León had observed) that was assumed to have served as a fortress; it is now known to have been more like a stage-pyramid with built-in reservoirs, conduits, channels and sluices that indicate its true purpose: a facility for the separation and processing of ores.

This artificial hill, which some believe originally had the shape of a step-pyramid like that of a Mesopotamian ziggurat, dominates the flat landscape. As the visitor casts his gaze about, another structure stands out. Situated to the northwest of the Akapana, it appears from a distance as a transplanted *Arc de Triomphe* from Paris. It is indeed a gate, intricately cut and carved out of a single cyclopean stone block; but it was not set up to commemorate a victory—rather, to enshrine in stone a marvelous calendar.

Called "Gate of the Sun," the single stone block from which it was cut and shaped measured about ten by twenty feet and weighed more than one hundred tons. There are niches and geometrically accurate cutouts upon the lower part of the gate, especially on what is considered its back side (Fig. 122b). The most elaborate and enigmatic carvings are on the upper front side (Fig. 122a), facing due east. There, the arch of the gate has been carved to depict in relief a central figure—probably of Viracocha—flanked on each side by three rows of winged attendants (Fig. 123a); the central figure and three rows have been positioned above a meandering geometric frame so carved as to snake over and under miniature images of Viracocha (Fig. 123b).

The writings of Posnansky have established that the carvings on the gate represented a twelve-month calendar of a year beginning on the day of the spring equinox in the southern hemisphere (September), yet a year where the other major points of the solar year—the autumn equinox and the two solstices—are also indicated by the positions and shapes of the depicted smaller images. It was, he concluded, a calendar of eleven months of thirty days each plus a "great month," a twelfth month of thirty-five days, adding up to a solar year of 365 days.

A twelve-month year beginning on the day of the spring equinox was, as we now know, first introduced at Nippur, in Sumer, circa 3800 B.C.

The "Gate of the Sun," archaeologists have discovered, stands at the northwest corner of what was a wall constructed of upright stone pillars that formed a rectangular enclosure within which the third most prominent edifice of the site

Figures 122a and 122b

stood. Some believe that there was originally a similar gate at the southwestern corner of the enclosure, flanking symmetrically a row of thirteen monoliths erected in the precise center of the enclosure's western wall. That row of monoliths, part of a special platform, faced exactly the monumental stairway that was built at the center of the eastern wall, on the enclosure's opposite side. The monumental stairway, which has been unearthed and restored, led to a series of raised rectangular platforms that encompassed a sunken courtyard (Fig. 124a).

Given the name *Kalasasaya* ("The Standing Pillars"), the edifice was thus oriented precisely along an east–west

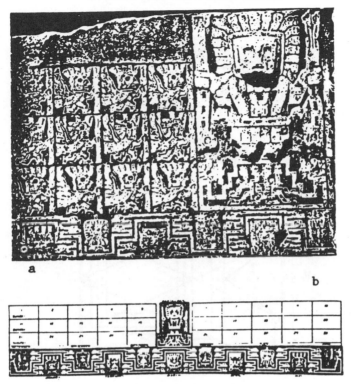

Figures 123a and 123b

axis, in the manner of the Near Eastern temples. This was the first clue that it could have served astronomical purposes. Subsequent researches indeed established that it was a sophisticated observatory for determining the solstices as well as the equinoxes by observing sunrises and sunsets from certain focal points along sightlines anchored at the enclosure's corners and the pillars erected at its western and eastern walls (Fig. 124b). Posnansky found evidence that the back side of the Gate of the Sun was so carved out that it probably held two golden panels that could be swung on bronze axles; that might have enabled the astronomer-priests to angle the plates so that they reflected the sunset rays toward any desired observation post in the Kalasasaya proper. The multiple sightlines, more than were required just for observations on solstice or equinox days, the fact that Viracocha was helped by both the Sun and the Moon,

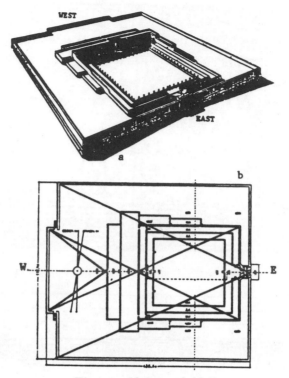

Figures 124a and 124b

and the fact that there were thirteen, not just twelve, pillars at the center of the western wall suggest that the Kalasasaya was not just a solar observatory, but one that served a solar-lunar calendar.

The realization that this ancient structure, more than twenty thousand feet up the Andean mountains, in a desolate, narrow plain among snowbound mountains, was a sophisticated calendrical observatory was compounded by discoveries regarding its age. Posnansky was the first to conclude that the angles formed by the lines of sight suggested an obliquity somewhat greater than the present declination of 23.5°; it meant, he himself was astounded to realize, that the Kalasasaya had been designed and built thousands of years before the Common Era.

The understandable disbelief on the part of the scientific community at the time—at most it was thought the ruins, if not from Incan times, were not older than from a few centuries B.C.—led to the dispatch of a German Astronomical Commission to Peru and Bolivia. Dr. Rolf Müller, whose extensive work on other sites has already been mentioned by us, was one of the three astronomers chosen for the task. The investigations and thorough measurements left no doubt that the obliquity prevailing at the time of construction was such that the Kalasasaya could have been built circa 4050 B.C. or (as the Earth tilted back and forth) circa 10,050 B.C. Müller, who had arrived at a date of just over 4000 B.C. for the megalithic remains at Machu Picchu, was inclined likewise to date the Kalasasaya—a conclusion with which Posnansky in the end agreed.

Who was there with such sophisticated knowledge to plan, orient, and erect such calendrical observatories—and in a manner that followed the astronomical principles and calendrical arrangements devised in the ancient Near East? In *The Lost Realms* we presented the evidence and arrived at the conclusion that it was the same Anunnaki, those who had come to Earth from Nibiru in need of gold. And, like the men who searched for the golden El Dorado millennia later, they also came to the New World in search of gold. The mines in southeastern Africa were flooded by the Deluge; but the same upheaval uncovered the incredibly rich veins of gold in the Andes.

We believe that Anu and his spouse Antu, visiting Earth from Nibiru circa 3800 B.C., also went to see for themselves the new metallurgical center on the southern shore of Lake Titicaca. They left by sailing away on the lake from the port facilities of Puma Punku, where the cyclopean chambers, carved and shaped out of single stone blocks, then stood alongside massive piers.

The remains at Puma Punku hold another enigmatic clue to the amazing link between the structures at Lake Titicaca and the unusual temple to Ninurta that Gudea built. To the disbelief of the site's excavators, they found that the megalithic builders had used *bronze clamps,* formed to fit T-shaped cutouts in adjoining stones, to hold together the huge stone blocks (Fig. 125). Such a clamping method, and such

Figure 125

a use of bronze, were unique to the Megalithic Age, having been found only at Puma Punku and at another site of cyclopean megaliths, Ollantaytambu, some forty-five miles northwest of Cuzco in the Sacred Valley.

Yet thousands of miles away, on the other side of the world, at Lagash in Sumer, Gudea used the very same unique method and the very same unique bronze clamps to hold together the stones that, imported from afar, were used in the construction of the Eninnu. Recording in his inscriptions the unusual use of stones and of metals, this is how Gudea lauded his own achievements:

> *He built the Eninnu with stone,*
> *he made it bright with jewels;*
> *with copper mixed with tin [bronze]*
> *he held it fast.*

It was a feat for which a *Sangu Simug*, a "priestly smith," was brought over from the "Land of Smelting." It was, we believe, Tiahuanacu in the Andes.

10

IN THEIR FOOTSTEPS

The Great Sphinx of Egypt gazes precisely eastward, welcoming the rising Sun along the 30th parallel. In ancient times its gaze welcomed the Anunnaki "gods" as they landed at their spaceport in the Sinai peninsula, and later on guided the deceased pharaohs to an Afterlife, when their *Ka* joined the gods in their heavenly ascents. At some time in between, the Sphinx might have witnessed the departure of a great god—Thoth—with his followers, to be counted among the First Americans.

The 500th anniversary of the epochal voyage of Columbus in 1492 has been by now reclassified from discovery to rediscovery, and has intensified the inquiry regarding the true identity of the "First Americans." The notion that settlement of the Americas began with the trekking of family groups from Asia across a frozen landbridge to Alaska, just before the last ice age abruptly ended, has been grudgingly giving way in the face of mounting archaeological evidence that humans arrived in the Americas many millennia earlier, and that South America, not North America, was the earliest arena of human presence in the New World.

"For the last 50 years, the received wisdom has been that the 11,500-year-old artifacts found at Clovis, New Mexico, were made soon after the first Americans found their way across the Bering landbridge," *Science* magazine (21 February 1992 issue) wrote in an update on the debate among scientists; "Those who have dared question the consensus have met with harsh criticism." The reluctance to accept an earlier age and a different arrival route stems primarily from the simple assumption that Man could not have crossed the oceans separating the Old and New Worlds at such prehistoric times because maritime technology did not yet exist. Notwithstanding the evidence to the contrary, the rock-bottom logic continues to be, if Man couldn't do it, it didn't happen.

The age of the Sphinx has recently emerged as an analogous issue, where scientists refuse to accept new evidence because it implies achievements by Man when Man could not have achieved them; and guidance or assistance by the "gods"—Extraterrestrials—is simply out of consideration.

In previous books of *The Earth Chronicles* we have presented extensive evidence (to date unrefuted) that the great pyramids of Giza were built not by pharaohs of the Fourth Dynasty circa 2600 B.C., but by the Anunnaki "gods" millennia earlier, as components of the landing corridor for the spaceport in the Sinai peninsula. We arrived at the time frame of circa 10,000 B.C.—some 12,000 years ago—for those pyramids; and we showed that the Sphinx, built soon thereafter, had already existed on the Giza plateau when pharaonic reigns began many centuries before the Fourth Dynasty. The evidence we relied on and presented were Sumerian and Egyptian depictions, inscriptions, and texts.

In October 1991, some fifteen years after our initial presentation of such evidence in *The 12th Planet,* Dr. Robert M. Schoch, a Boston University geologist, reported at the annual meeting of the Geological Society of America that meteorological studies of the Sphinx and its layering indicated that it was carved out of the native rock "long before the dynasties of the Pharaohs." The research methods included seismic surveying of subsurface rocks by Dr. Thomas L. Dobecki, a geophysicist from Houston, and Egyptologist Anthony West of New York, and the study of weathering and watermarks on the Sphinx and its surroundings. The precipitation-induced weathering, Dr. Schoch stated, "indicated that work on the Sphinx had begun in the period between 10,000 B.C. and 5000 B.C., when the Egyptian climate was wetter."

The conclusion "flies in the face of everything we know

about ancient Egypt,'' the *Los Angeles Times* added in its report of the announcement. ''Other Egyptologists who have looked at Mr. Schoch's work cannot explain the geological evidence, but they insist that the idea that the Sphinx is thousands of years older than they had thought just simply ''does not match up'' with what has been known. The newspaper quoted archaeologist Carol Redmount of the University of California at Berkeley: ''There's just no way that could be true . . . The Sphinx was created with technology that was far more advanced than that of other Egyptian monuments of known date, and the people of that region would not have had the technology, the governing institutions or the will to have built such a structure thousands of years earlier.''

In February 1992, the American Association for the Advancement of Science, meeting in Chicago, devoted a session to the subject ''How old is the Sphinx?'' at which Robert Schoch and Thomas Dobecki debated their findings with two debunkers, Mark Lehner of the University of Chicago and K.L. Gauri of the University of Louisville. According to the Associated Press, the heated debate which spilled over into a confrontation in the hallway has not focused on the scientific merits of the meteorological findings, but, as Mark Lehner expressed it, on whether it is permissible to ''overthrow Egyptian history based on one phenomenon, like a weathering profile.'' The final argument by the debunkers was the absence of evidence that a civilization advanced enough to carve the Great Sphinx existed in Egypt between 7000 and 5000 B.C. ''The people during that age were hunters and gatherers; they didn't build cities,'' Dr. Lehner said; and with that the debate ended.

The only response to this logical argument is, of course, to invoke someone other than the ''hunters and gatherers'' of that era—the Anunnaki. But admitting that all evidence points to such more advanced beings from another planet is a threshold that not everyone, including those who find the Sphinx to be 9,000 years old, is as yet ready to cross.

The same Fear-of-Crossing (to coin an expression) has

blocked for many years not just the acceptance, but even the dissemination, of evidence concerning the antiquity of Man and his civilizations in the Americas.

The discovery near Clovis, New Mexico, in 1932 of a trove of leaf-shaped, sharp-edged stone points that could be attached to spears and clubs for hunting, and subsequently at other North American sites, led to the theory that big game hunters migrated from Asia to the Pacific northwest some 12,000 years ago, when Siberia and Asia were linked by an icy landbridge. In time, the theory held, these "Clovis People" and their kindred folk spread over North America and, via Central America, eventually also to South America.

This neat image of the First Americans retained its exclusive hold in spite of occasional discoveries, even in the southwestern United States, of remains of crushed bones or chipped pebbles—arguably evidence of human presence—dating some 20,000 years before Clovis. A less doubtful find has been that at Meadowcroft rock shelter, Pennsylvania, where stone tools, animal bones, and, most important, charcoal, have been carbon-dated to between 15,000 and 19,000 years ago—millennia before Clovis, and in the eastern part of the United States to boot.

As linguistic research and genetic trace-backs joined other investigative tools, the evidence began to mount in the 1980s that humans arrived in the New World some 30,000 years ago—probably in more than one migration, and perhaps not necessarily over an icebridge but by rafts or canoes hugging the coastlines. The basic tenet—out of northeast Asia into northwest America—has, however, been stubbornly maintained in spite of unsettling evidence from South America. That evidence, whose discovery was not only ignored but even initially suppressed, pertains primarily to two sites where Stone Age tools, crushed animal bones, and even petroglyphs have been found.

The first of these unsettling settlement sites is Monte Verde in Chile, on the continent's Pacific side. There archaeologists have found remains of clay-lined hearths, stone tools, bone implements, and foundations of wooden shelters—a campsite occupied some 13,000 years ago. This is a date much too early to be explained by a slow southward migration of Clovis People from North America. Moreover, lower strata at this campsite yielded fragmented stone tools that suggest that the site's human occupation began some 20,000 years earlier. The second site is all the way on the other side of South America, in Brazil's northeast. At a place called Pedra Furada, a rock shelter contained circular hearths filled with charcoal surrounded by flints; the nearest source of flint is a mile away, indicating that the sharp stones were brought over intentionally. Dating by radiocarbon and newer methods provided readings spanning the period 14,300 to 47,000 years ago. While most established archaeologists continue to consider the early dates "simply inconceivable," the rock shelter has yielded, at the 10,000 B.C. level, petrogylphs (rock paintings) whose age is undisputable. In one, a long-necked animal that looks like a giraffe—an animal nonexistent in the Americas—seems to have been depicted.

The ongoing challenge to the Clovis theory in regard to the time of arrival has been accompanied by a challenge to the via-the-Bering-strait route as the sole path of arrival. Anthropologists at the Arctic Research Center of the Smithsonian Institution in Washington, D.C., have concluded that the image of animal-skin clad hunters carrying spears across a frozen wilderness (with women and children in tow) is all wrong in thinking of the First Americans. Rather, they were maritime people who sailed in rafts or skinboats to the more hospitable southern shores of the Americas. Others, at the Center for the Study of the First Americans at Oregon State University, do not rule out a crossing of the Pacific via the islands and Australia (which was settled circa 40,000 years ago).

Most others still consider such early crossings by "primitive man" as fantasies; the early dates are shrugged off as instrumental errors, stone "tools" as pieces of fallen rocks, broken animal bones as crushed by rockfalls, not by hunters. The same question that has brought the Age of the Sphinx debate to a dead end has been applied to the First Americans debate: who was there, tens of thousands of years ago, who possessed the technology required for crossing vast oceans by boat, and how could those prehistoric mariners have known that there was land, habitable land, on the other side?

This is a question that (also when applied to the Age of the Sphinx) has only one answer: the Anunnaki, showing Man how to cross the oceans, telling him why and where-to—perhaps carrying him over, "on the wings of eagles," as the Bible has described—to a new Promised Land.

There are two instances of planned migrations recounted in the Bible, and in both the deity was the guide. The first instance was the ordering of Abraham, more than 4,000 years ago, to "get thee out of thy country and out of thy birthplace and from thy father's house." He was to go, Yahweh said, "unto the land which I will show thee." The second instance was the Israelite Exodus from Egypt, some 3,400 years ago. Showing the Israelites the route to take to the Promised Land,

> *Yahweh went before them by day*
> *in a pillar of cloud,*
> *to lead them the way,*
> *and by night in a pillar of fire*
> *to give light to them,*
> *to go by day and by night.*

Aided and guided, the people followed in the footsteps of the gods—in the ancient Near East as well as in the new lands across the oceans.

The latest archaeological discoveries lend credence to memories of early events that are called "myths" and "legends." Invariably, they speak of multiple migrations and always from across the seas. Significantly, they often involve the numbers seven and twelve—numbers that are not a reflection of human anatomy or digital counting, but a clue to astronomical and calendrical knowledge, as well as to links with the Old World.

One of the best preserved cycle of legends is that of the Nahuatl tribes of central Mexico, of which the Aztecs whom the Spaniards encountered were the latest extant. Their tales of migration encompassed four ages, or "Suns," the first one of which ended with the Deluge; one version that provides lengths in years for those ages indicates that the first "Sun" began 17,141 years before the tale was related to the Spaniards, i.e. circa 15,600 B.C. and thus indeed millennia before the Deluge. The earliest tribes, the oral legends and the tales written down pictorially in books called codices related, came from *Azt-lan,* the "White Place," which was associated with the number seven. It was sometimes depicted as a place with seven caves out of which the ancestors had emerged; alternatively, it was painted as a place with seven temples: a central large step-pyramid (ziggurat) surrounded by six lesser shrines. *Codex Boturini* contains a series of cartoonlike paintings of the early migration by four tribes that began from the place of the seven temples, involved crossing a sea in boats and a landing in a place of cave shelters; the migrants were guided in that journey to the unknown by a god whose symbol was a kind of Seeing-eye attached to an elliptical rod (Fig. 126a). The four clans of migrants then trekked inland (Fig. 126b), passing by and following various landmarks. Splitting into several tribes, one, the *Mexica,* finally reached the valley where an eagle was perched upon a cactus bush—the signal for their final destination and the place where the Nahuatlan capital was to be built. It later developed into the Aztec capital, whose symbol remained the eagle perched on a cactus bush. It was called *Tenochtitlan,* the City of Tenoch. Those earliest migrants were called Tenochites, the People of Tenoch; in *The Lost Realms* we detailed the reasons why they might have been the descendants of Enoch, the son of Cain, who still suffered the sevenfold avenging of their forefather's crime of fratricide. According to the Bible Cain, who was banished to a distant "Land of Wandering," built a city and named it after his son Enoch; and Enoch had four descendants from whom there grew four clans.

The Spanish chronicler Friar Bernardino de Sahagún

Figures 126a and 126b

(*Historia de las cosas de la Nueva Espana*), whose sources were verbal as well as Nahuatlan tales written down after the conquest, recorded the sea voyage and the name, *Panotlan,* of the landing site; the name simply meant "Place of arrival by sea," and he concluded that it was in what is now Guatemala. His information added the interesting detail that the immigrants were led by four Wise Men, "who carried with them ritual manuscripts and who also knew the secrets of the calendar." We now know that the two—ritual and calendar—were two sides of the same coin, the worship of the gods. It is a safe bet that the Nahuatlan calendar followed the twelve-month arrangement, perhaps even the twelve-zodiac division; for we read (in Sahagún's chronicles) that the Toltecs, the Nahuatl tribe that preceded and taught the Aztecs, "knew that many are the heavens; they said that there are twelve superimposed divisions" thereof.

Down south, where the Pacific Ocean lapped the coasts of South America, Andean "myths" did not recall pre-Diluvial migrations but knew of the Deluge and asserted that the gods, already present in those lands, were the ones to help the few survivors upon the high peaks to repopulate the continent. The legends do speak clearly of new, post-Diluvial arrivals by sea; the first or most memorable of them was one headed by a leader called Naymlap. He led his people across the Pacific in a fleet of boats made of balsa wood, guided by an "idol," a green stone through which the Great God delivered navigational and other instructions. The landfall was at the point where the South American continent juts out the most westward into the Pacific Ocean, at what is nowadays called Cape Santa Helena in Ecuador. After they had landed, the Great God (still speaking through the green stone) instructed the people in farming, building, and handicrafts.

An ancient relic made of pure gold, now kept in the Gold Museum of Bogotá, Colombia (Fig. 127), depicts a tall leader with his entourage atop a balsa wood raft. The artwork may well have represented the sea crossing by Naymlap or his like. They were well acquainted, according to the Naymlap legend, with the calendar and worshiped a pan-

Figure 127

theon of twelve gods. Moving inland to settle where Quito, Ecuador's capital, is now situated, they built there two temples facing each other: one dedicated to the Sun, the other to the Moon. The Temple of the Sun had in front of its gateway two stone columns and in its forecourt a circle of twelve stone pillars.

The familiarity with the sacred number twelve—the hallmark of the Mesopotamian pantheon and calendar—bespeaks a calendar not unlike the one that originated in Sumer. The veneration of both the Sun and the Moon indicates a solar-lunar calendar, again as the one begun in Sumer. A gateway with two stone columns in front of it brings to mind the two columns that were erected at the entrances to temples throughout the ancient Near East, from Mesopotamia through western Asia and Egypt. And, as if all those links to the Old World were not enough, we find a *circle of twelve stone pillars*. Whoever had arrived from across the Pacific must have been aware of the astronomical stone circles of Lagash, or Stonehenge— or both.

Several stone objects that are now kept in the National Museum of Peru in Lima are believed to have served the coastal peoples as calendrical computers. One, for example, catalogued under the number 15-278 (Fig. 128) is divided into sixteen squares that contain pegholes that range from six to twelve; the top and bottom panels are indented with twenty-nine and twenty-eight pegholes respectively — a strong suggestion of a count of lunar monthly phases.

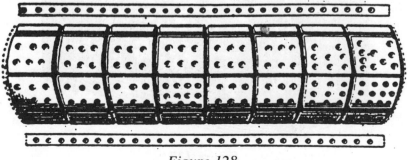

Figure 128

Fritz Buck (*Inscriptiones Calendarias del Peru Preincaico*) who made the subject his specialty, was of the opinion that the 116 pegholes or indentations in the sixteen squares indicated a link to the calendar of the Mayas of Mexico and Guatemala. That the northern parts of the Andean lands were in close contact with the people and cultures of Mesoamerica—a possibility until recently rejected out of hand—is now hardly disputed. Those who arrived from Mesoamerica undoubtedly included African and Semitic people, as evidenced by numerous stone carvings and sculptures (Fig. 129a). Before them there arrived by sea people that were depicted as Indo-Europeans (Fig. 129b); and sometime in between there landed on these coasts helmeted "Bird people" (Fig. 129c) who were armed with metal weapons. Another group may have arrived overland via the Amazon basin and its tributaries;

Figures 129a, 129b, and 129c

the symbols that were associated with them (Fig. 130) were identical to the Hittite hieroglyph for "gods." Inasmuch as the Hittite pantheon was an adaptation of the Sumerian pantheon, it perhaps explains the otherwise remarkable discovery of a golden statuette in Colombia of a goddess holding in her hands the emblem of the umbilical cutter—the emblem of Ninharsag, the Mother Goddess of the Sumerians (Fig. 131).

The north-central Andean coast and ranges of South America were peopled by Quechua-speaking peoples, named, for want of a better source, after the main rivers along which they flourished. The Incas, it turned out, formed their empire and laid out their famous highways upon the ruins of those earlier inhabitants. Down south, from about where Lima (the capital of Peru) is situated, along the coast and mountains that face Lake Titicaca, and on southward

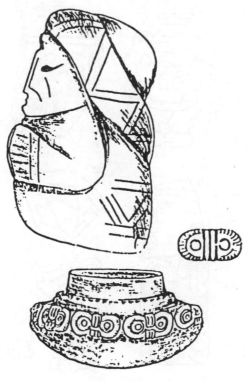

Figure 130

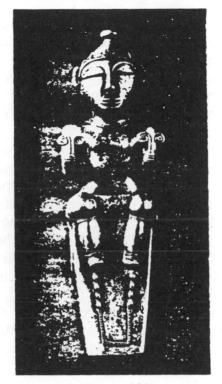

Figure 131

toward Chile, the dominant tribal language was that of the Aymaras. They too recalled in their legends early arrivals, on the Pacific coast by sea and by land from the territory east of Lake Titicaca. The Aymara considered the former as unfriendly invaders; the latter were called *Uru,* meaning "Olden people," who were a people apart and whose remnants still exist in the Sacred Valley as a group with its own customs and traditions. The possibility that they were Sumerians, arriving at Lake Titicaca when Ur was Sumer's capital (the last time between 2200 and 2000 B.C.), must be taken seriously. The fact is that the province that connects the Sacred Valley, the eastern shores of Lake Titicaca, and western Brazil is still called *Madre del Dios*—"Mother of the gods," which is what Ninharsag was. A mere coincidence?

Scholars find that throughout the millennia the dominant

cultural influence on all these peoples was that of Tiahuanacu; it found its most obvious expression in the thousands
of clay and metal objects that bore the image of Viracocha
as it appears on the Gate of the Sun, in decorations (including on the magnificently woven cloth in which mummies were wrapped) that emulated the symbols on the Gate,
and in their calendar.

The most prevalent of those symbols or, as Posnansky
and others consider them, hieroglyphs, was that of the stairway (Fig. 132a), which was also used in Egypt (Fig. 132b)
and which was often used on Andean artifacts to denote a
"Seeing-eye" tower (Fig. 132c). Such observations, to
judge from the astronomical lines of sight at the Kalasasaya
and from the celestial symbols associated with Tiahuanacu,
included the Moon (whose symbol was a circle between
crescents, Fig. 132d).

On the Pacific side of South America, it thus appears,
the calendar and its celestial knowledge followed in the
footsteps of the same teachers who had been active in the
Near East.

Commenting on the evidence, earlier discussed, for the
much greater antiquity of human presence in the Americas

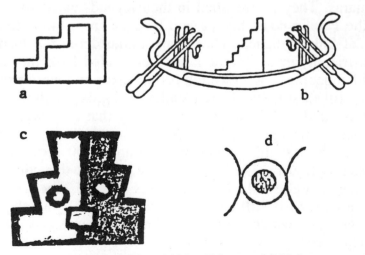

Figures 132a, 132b, 132c, and 132d

and their routes of arrival, Dr. Niede Guidon, of the French Institute of Advanced Social Studies who participated with Brazilian archaeologists in the Pedra Furada discoveries, said thus: "A transatlantic crossing from Africa cannot be ruled out."

The discovery of "the oldest pottery in the Americas," announced by an archaeological team of the Field Museum of Natural History in Chicago in the December 13, 1991, issue of *Science* magazine, "overturned the standard assumptions" regarding the peopling of the Americas and especially the view that the Amazon basin, where the discovery was made, was "simply too poor in resources to have supported a complex prehistoric culture." Contrary to long held opinions, "the Amazon basin had soil as fertile as the flood plains of the Nile, the Ganges and other great river basins of the world," said Dr. Anne C. Roosevelt, the team's leader. The red-brown pottery fragments, some decorated with painted patterns, have definitely been dated by the latest technologies to be no less than seven thousand years old. They were found at a site called Sabtarem in mounds of shells and other trash discarded by the ancient residents, a fishing people.

The date, and the fact that the pottery was painted with linear designs, put it on a par with similar pottery that appeared in the ancient Near East, in the mountains bordering on the plain where the Sumerian civilization blossomed out. In *The Lost Realms* we presented the evidence of Sumerian traces in the Amazon basin and through it in the gold- and tin-producing areas of Peru. The latest discovery, by fixing the pottery's date unquestionably and by coming at a time when early arrivals are a more acceptable possibility, serves mainly to corroborate previously unorthodox conclusions: in antiquity, people from the Near East reached America also by crossing the Atlantic Ocean.

Arrivals from such a direction have not been without calendrical remains. The most dramatic and enigmatic of them were discovered in the northeastern part of the Amazon basin, near the Brazil-Guyana border. There, rising in the great plain, is an egg-shaped rock that rises some 100 feet and is some 300 by 250 feet in diameter. A natural cavity on its top has been carved out to form a pond whose waters flow on and into the gigantic rock through channels and conduits. A cavelike cavity has been enlarged to form a large rock shelter, further carved out to form grottoes and platforms at various levels. The entrance into the rock's innards has painted above it a snake that is about twenty-two feet long, its mouth formed of three openings into the rock that are surrounded by enigmatic and undeciphered inscriptions; inside and out, the rock is filled with hundreds of painted signs and symbols.

Intrigued by reports by earlier explorers and local lore that the grottoes contained skeletons of ''giants whose faces were European in expression,'' Professor Marcel F. Homet (*Die Söhne der Sonne*) explored the rock in the 1950s and provided more accurate data about it than had been known. He found that the three facades of the Pedra Pintada point in three directions: the large facade is oriented on an east–west line, and the two smaller ones are oriented south-southeast and south-southwest. His observation was that ''Externally, in its structural orientation . . . this monument follows the exact identical rules of the ancient European and Mediterranean cultures.'' He considered many of the signs and symbols painted on the meticulously polished surfaces of the rock to be ''impeccably regular numerals which are not based on the decimal system'' but ''belong to the oldest known eastern Mediterranean cultures.'' He thought that surfaces filled with dots represented tables of multiplication, such as 9 times 7 or 5 times 7 or 7 times 7, and 12 times 12.

The highlight of the rock's ancient artifacts, because of which some earlier explorers had called it the Place of the Stone Books, were dolmens—large flat stones laid across supporting stones—weighing between fifteen and twenty tons each. They were elaborately painted on their faces; and two larger ones were cut into precise shapes— one as a pentagon (Fig. 133a) and the other as an oval (Fig. 133b). As at the entrance, both appear to depict a

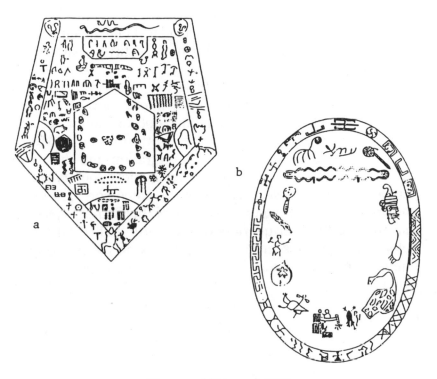

Figures 133a and 133b

serpent as the dominant symbol, and this and other signs brought to Homet's mind ancient Egypt and the eastern Mediterranean. Since many of the dolmens were placed at the levels and the entrances of burial grottoes in the depths of the rock, he concluded that, as the Indian legends held, this was a sacred place for the burial of leaders or other notables "by civilized people who were here, just as they were in Tiahuanacu, the great city of the Andes long, long ago—perhaps thousands of years before the birth of Christ."

Homet's observation regarding the mathematical system that seemed to underlie the markings on the surfaces, "not based on the decimal system" but on that of "the oldest known eastern Mediterranean cultures," is a roundabout way of describing the Sumerian sexagesimal system whose use prevailed throughout the ancient Near East. His other conclusions about links on the one hand to the "eastern Mediterranean" and on the other hand to Tiahuanacu "thousands of years before the birth of Christ" are truly remarkable.

Although the drawings on these two particular dolmens remain undeciphered, they do hold, in our view, a number of important clues. The pentagonal one no doubt records some coherent tale, perhaps, as with the later Mesoamerican picture books, a tale of migration and the route taken. At its four corners the tablet depicts four types of people; in that it could have been a precursor of a well-known Mayan painting on the cover of the *Codex Fejérvary* that showed the four quarters of the Earth and (in different colors) their diverse races of people. As on the pentagonal dolmen, the Mayan depiction also has a geometric central panel.

Except for the central panel, which in Brazil is pentagonal, the dolmen's face is covered with what appears to be an unknown script. We find similarities between it and a script from the eastern Mediterranean known as Linear A; it was a precursor of the script of the island of Crete, and also of that of the Hittites of Anatolia (today's Turkey).

The dominant symbol on the pentagonal dolmen is the serpent, also a well-known symbol of the pre-Hellenic culture of Crete and ancient Egypt. In terms of the ancient Near Eastern pantheon, the serpent was the symbol of Enki and his clan. On the oval dolmen it is depicted as a heavenly cloud, which brings to mind the the serpent symbol on Mesopotamian *kudurru* (Fig. 92), where it represented the Milky Way.

Many of the symbols that frame the central panel on this dolmen are familiar Sumerian and Elamite designs and emblems (such as the swastika). The larger images within the oval frame are even more revealing. If we consider the central uppermost symbol as a script element, precisely twelve symbols are left. In our view, they represent the *twelve signs of the zodiac*.

That not all the symbols are identical to those that originated in Sumer is not unusual, since in various lands (such as China) the zodiac (which means "animal circle") was adapted to local fauna. But some of the symbols on this oval dolmen, such as that of the two fishes (for Pisces), the two human images (the twins of Gemini) and the female holding a stalk of grain (Virgo, the Virgin) are identical to the zodiacal symbols (and their names) that originated in Sumer and were adopted throughout the Old World.

The significance of the Amazonian depiction can, therefore, hardly be exaggerated. As we have pointed out, the zodiac was an entirely arbitrary division of the celestial circle into twelve groups of stars; it was not the result of simple observation of natural phenomena, such as the day-night cycle, the waxing and waning of the Moon, or the Sun's seasonal changes. To find the concept and knowledge of the zodiac, and moreover to have it represented by Mesopotamian symbols, must be taken as evidence of someone with Near Eastern knowledge in the Amazon basin.

No less astounding than the decorative symbols and the zodiacal signs around the oval dolmen's face is the depiction in the center of the pentagonal dolmen. It shows a *circle of stones* surrounding two monoliths, between which there appears a partly erased drawing of a human head whose eye is focused on one of the monoliths. Such a "head with sighting eye" can be found in Mayan astronomical codices, in which the sign depicts astronomer-priests.

All that, plus the astronomical orientations of the rock's three surfaces, bespeaks the presence of someone familiar with celestial observations.

Who was that "someone"? Who could have crossed the ocean at such an early time? The crossing, admittedly, could not have taken place unaided. And whether those who were led or transported to the South American shores already possessed calendrical-astronomical knowledge, or were taught it in the new lands, none of that could have come about without the "gods."

* * *

In the absence of written records, the petroglyphs that have been found in South America are precious clues to what the ancient inhabitants had known and seen. Many of them have been found in the funnel leading, in the continent's northeastern part, into the Amazon basin and up that mighty river and its countless tributaries that begin in the distant Andes. The principal river of the Sacred Valley of the Incas, the Urubamba, is but a tributary of the Amazon River; so are other Peruvian rivers that flow eastward from sites whose mind-boggling remains indicate they were metallurgical processing centers. The known sites, only a fraction of what is there to be discovered if proper archaeological work were carried out, support the veracity of local traditions that people from across the Atlantic landed on those coasts and journeyed via the Amazon basin to obtain the gold and tin and other treasures of the Andes.

In what used to be called British Guiana alone, more than a dozen sites have been discovered where the rocks are covered with carved pictures. At a site near Karakananc in the Pacaraima mountains, the petroglyphs (Fig. 134a) depict stars with different numbers of rays or points (a Sumerian "first"), the crescent of the Moon and solar symbols, and what could have been a viewing device next to a stairway. At a place called Marlissa a long range of granite rocks along a river bank is covered with numerous petroglyphs; some of them adorned the cover of the journal of the Royal Agricultural and Commercial Society of British Guiana (*Timehri*, issue 6 of 1919) (Fig. 134b). The peculiar person with raised hands and a helmetlike head with one large "eye" appears on the rock next to what looks like a large boat (Fig. 134c). The tightly clothed and haloed beings, shown many times over (Fig. 134d), are of giantlike proportions: in one instance thirteen feet tall and in another close to eight feet.

In neighboring Suriname, formerly Dutch Guiana, in the area of the Frederik Willem IV Falls, the petroglyphs are so numerous that researchers have found it necessary to assign numbers to the sites, to each group of petroglyphs

Figures 134a, 134b, 134c, and 134d

at each site, and to individual symbols within each group. Some of them (Fig. 135) would today be deemed to represent UFOs and their occupants, as would a petroglyph (Fig. 136) at site 13 at the Wonotobo Falls, where the previously seen depiction of tall and haloed beings has been converted into a domed contraption with a ladder coming down out of its opening; a mighty person is standing in that opening.

The message conveyed by such petroglyphs is that while some people were seen arriving by boats, other godlike ones arrived in "flying saucers."

At least two of the symbols among these petroglyphs can be recognized as Near Eastern script-signs, and specifically so from Hittite inscriptions in Anatolia. One, which appears as the determinative-sign next to a helmeted and horned face (Fig. 137a), unmistakably resembles the hieroglyphic Hittite

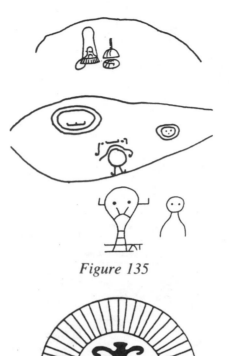

Figure 135

Figure 136

sign meaning "great" (Fig. 137b). This hieroglyphic sign was most often used in Hittite inscriptions in combination with the sign for "king, ruler" to mean "great king" (Fig. 137c); and exactly such a combined hieroglyph has been found several times among the petroglyphs near the Wonotobo cataracts in Suriname (Fig. 137d).

Petroglyphs, indeed, cover rocks large and small throughout South America; their spread and images tell Man's story in that part of the world, a story that is yet to be fully deciphered and understood. For more than a hundred years explorers have shown that the South American continent

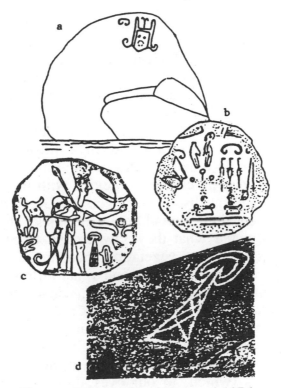

Figures 137a, 137b, 137c, and 137d

can be crossed by foot, on horseback, by canoes and rafts. One major route begins in northeastern Brazil/Guyana/Venezuela and uses principally the Amazon river system to enter Peru's north and central parts; the other begins in Brazil somewhere near São Paulo and winds its way westward through the Mato Grosso region to Bolivia and Lake Titicaca, and thence northward into either central Peru (the Sacred Valley) or the coastal regions—two places where the two routes meet.

As the discoveries earlier discussed in this chapter show, Man arrived in the Americas, and especially in South America, tens of thousands of years ago. The migrations, to judge by the petroglyphic evidence, came in three recognizable phases. The extensive work at the Pedra Furada in northeastern Brazil offers a good example of those phases as far as the continent's Atlantic side is concerned.

Pedra Furada is just the most studied site in the area named after its principal village, São Raimundo Nonato; more than 260 archaeological sites of early occupation are found there, and 240 of them contain rock art. As the carbon dating of the charcoal samples from the prehistoric hearths shows, Man lived there beginning some 32,000 years ago. Throughout the area, such habitation appears to have come to an abrupt end circa 12,000 years ago, concurrently with a marked change in climate. It has been our opinion that the change coincided with the abrupt end of the last ice age by the Deluge, the Great Flood. The rock art of that long period was naturalistic; the artists of the time depicted what they saw around them: local animals, trees and other vegetation, people.

A hiatus of some two thousand years followed until human occupation of the site resumed, when other and new groups arrived in the area. Their rock art suggests that they had come from a distant land, for animals not native to the area were included in the paintings: giant sloths, horses, an early type of llama, and (according to the excavators' reports) camels (which, to our eyes, looked more like giraffes). This second phase lasted till about 5,000 years ago and included, in its latter part, the making of decorated pottery. It also included in its art, in the words of Niede Guidon who has led the excavations, "abstract signs" that "seem related to ceremonies or mythical subjects"—a religion, an awareness of the "gods." It is at the end of that phase that the transition to petroglyphs akin to Near Eastern signs, symbols, and script make their appearance, leading in such a third phase to the astronomical and calendrical aspects of the markings on the rocks.

These petroglyphs can be found at both landfall zones and along the two major cross-continent routes. The more they belong to the third phase, the more pronounced are the celestial symbols and connotations. The more they are found in the continent's southern parts, be it in Brazil or Bolivia or Peru, the more are they reminiscent of Sumer, Mesopotamia, and Anatolia. Some scholars, especially in South America, interpret various signs as a kind of cuneiform Sumerian script. The largest petroglyph in that zone is the so-called candelabra or trident that faces whoever reaches South America's Pacific shore at the Bay of Paracas (Fig. 138a). According to local lore it is the lightning rod of Viracocha, as seen atop the Gate of the Sun in Tiahuanacu; we have identified it as the Near Eastern emblem of the "Storm God" (Fig. 138b), the younger son of Enlil whom the Sumerians called *Ishkur,* the Babylonians and Assyrians *Adad,* and the Hittites *Teshub* ("The Wind Blower").

While the Sumerian presence or at least influence can be documented in many, though small, ways, as we have done in *The Lost Realms,* no attempt has been made to date to arrive at a comprehensive picture of the Hittite presence in South America. We have shown some of the Hittite signs to be found in Brazil, but probably much more lies unearthed and unstudied behind such a coincidence as the fact that the hill people of Anatolia were the first to introduce iron in the Old World, and the parallel fact that the country's name, *Brazil,* is identical to the Akkadian word for iron, *Barzel*—a similarity that Cyrus

Figures 138a and 138b

H. Gordon (*Before Columbus* and *Riddles in History*) considered to be a significant clue regarding the true identity of early Americans. Other clues are the Indo-European types depicted on the busts found in Ecuador and northern Peru, and the fact that the enigmatic inscriptions found on Easter Island, in the Pacific Ocean opposite Chile, run as the Hittite script did in the "as the ox ploughs" system—beginning on the upper line from left to right, continuing on the second line from right to left, then again from left to right and so on.

Unlike Sumer, which was situated in an alluvial plain with no stones therein to serve as building materials, the Enlilite domain of Anatolia was all KUR.KI, "mountain land," of which Ishkur/Adad/Teshub was put in charge. The structures and edifices in the Andean lands were also made of stone—from the earliest cyclopean stoneworks through the exquisite ashlars of the Ancient Empire, down to the fieldstone buildings of the Incas and to the present. Who was there in the Andean lands knowledgeable in the use of stone for construction before the lands were populated, before Andean civilization began, before the Incas? We suggest they were stonemasons from Anatolia, who quite usefully were also expert miners—for Anatolia was an important source of metal ores in antiquity and one of the first places to begin mixing copper with tin to make bronze.

Making an on-site visit to the ruins of Hattusas, the ancient Hittite capital, and other bastions nearby, some 150 miles northeast of Ankara, the capital of present-day Turkey, one begins to realize that in some respects they represented crude emulations of Andean stoneworks, even including the unique and intricate incisions in the hard stone to create the "stairway motif" (Fig. 139).

One has to be an expert in ancient ceramics to be able to distinguish between some of the Anatolian and Andean pottery, especially the burnished and polished deep ocherred kind from the bronze age. One need not, however, be an expert to notice the similarity between strange warriors depicted on Peruvian artifacts from the coastal areas (Fig.

Figure 139

140a) and pre-Hellenic warriors depicted on artifacts from the eastern Mediterranean (Fig. 140b).

Regarding the latter similarity, it should be borne in mind that the home of the early Greeks, Ionia, was not in Greece but in the western parts of Anatolia (Asia Minor). The myths and legends of early times, recorded in such works as Homer's *Iliad*, deal in fact with locations that were in Anatolia. Troy was there and not in Greece. So was the famed Sardis, capital of Croesus, king of Lydia, who was renowned for his golden treasures. Perhaps the belief by some that the travels and travails of Odysseus also brought him to what we now call America, are not so farfetched.

Figures 140a and 140b

* * *

It is odd that in the increasingly heated debate about the First Americans, little if any attention has been given to the question of how much maritime knowledge the ancient peoples possessed. There are many indications that it was quite extensive and advanced; and once again, the impossible can be accepted as possible only if teachings by the Anunnaki are taken into account.

The Sumerian King List describes an early king of Erech, a predecessor of Gilgamesh, thus: "In Eanna, Meskiaggasher, the son of divine Utu, became high priest as well as king, and ruled 324 years. Meskiaggasher went into the western sea and came forth toward the mountains." How such a cross-oceanic voyage was accomplished without some kind of navigational aids, if none yet existed, is left unexplained by scholars.

Centuries later, Gilgamesh, having been mothered by a goddess, went in search of immortality. His adventures precede in time but exceed in drama those of Odysseus. On his last journey he had to cross the Waters or Sea of Death, which was possible only with the assistance of the boatman Urshanabi. No sooner did the two start the crossing, than Urshanabi accused Gilgamesh of breaking the "stone things'' without which the boatman could not navigate. The ancient text records the lament of Urshanabi about the ''broken stone things'' in three lines that are unfortunately only partly legible on the clay tablet; the three begin with the words ''I peer, but I cannot . . .'' which strongly suggests a navigational device. To correct the problem, Urshanabi instructed Gilgamesh to go back ashore and cut long wooden poles, 120 of them. As they sailed off, Urshanabi instructed Gilgamesh to discard one pole at a time, in groups of twelve. This was repeated ten times until all of the 120 poles were used up: ''At twice-sixty Gilgamesh had used up the poles,'' reaching their destination on the other side of the sea. Thus did a specific number of poles, arranged as instructed, substitute for the ''stone things'' that could no longer be used to peer with.

Gilgamesh is a known historical ruler of ancient Sumer; he reigned in Erech (Uruk) circa 2900 B.C. Centuries later, Sumerian traders reached distant lands by sea routes, exporting the grains, wool, and garments for which Sumer became known and importing—as Gudea has attested—metals, lumber, construction, and precious stones. Such two-way repeated voyages could not have taken place without navigational instruments.

That such instruments had existed in antiquity can be judged from an object that was found in the eastern Mediterranean off the Aegean island Antikythera at the beginning of this century. Sailing through the ancient sea route from the eastern to the western Mediterranean between the islands of Crete and Kythera, two boats of sponge divers discovered the wreck of an ancient ship lying on the sea's bottom. The wreck yielded artifacts, including marble and bronze statues, dated to the fourth century B.C. The ship itself has been dated to some time after 200 B.C.; amphorae, vessels containing wine, olive oil, and other foodstuffs, were dated to about 75 B.C. That the ship and its contents date to a time before the beginning of the Christian era thus seems certain, and so is the conclusion that it had taken on its load at or near the coast of Asia Minor.

The objects and materials raised from the wreck were taken to Athens for examination and study. Among them were a lump of bronze and broken-off pieces that, when cleaned and fitted together, stunned the museum officials. The ''object'' (Fig. 141) appeared to be a precise mechanism with many gears interlocked at various planes inside a circular frame that was in turn held in a square holder; it seemed to be an astrolabe ''with spherical projections and a set of rings.'' After decades-long studies, including its investigation with X rays and metallurgical analysis, it has been put on view in the National Archaeological Museum in Athens, Greece (catalog number X.15087). The protective housing bears a plaque that identifies the object as follows:

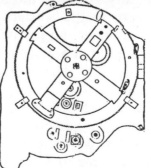

Figure 141

The mechanism was found in the sea of Antikythera island by sponge divers in 1900. It was part of the cargo of a shipwreck which occurred in the first century B.C.

The mechanism is considered to be a calendrical Sun and Moon computing machine dated, after the latest evidence, to circa 80 B.C.

One of the most thorough studies on the subject is the book *Gears from the Greeks* by Professor Derek de Sola Price of Yale University. He found that the three broken-apart sections contained gears and dials and graded plates that in turn were assembled from at least ten separate parts. The gears were linked one to the other on a basis of several differentials—a sophistication which we now find in automatic gearshift boxes in cars—that incorporated the cycle of the Sun and the Metonic (nineteen-year) cycle of the Moon. The gears were fitted with tiny teeth and moved on varied axles; markings on circular and angular parts were accompanied by inscriptions in Greek that named a number of zodiacal constellations.

The instrument was without doubt the product of a high technology and sophisticated scientific knowledge. Nothing coming even close to it in intricacy has been found in subsequent or preceding times, in spite of the guess offered by de Sola Price that it could have been made—or perhaps just repaired—at the School of Posidonios on the island of Rhodes after the model of planetarium devices used by Archimedes. Though he "sympathized with the shock one may feel at revising upwards the estimation of Hellenistic technology," he wrote, he could not agree with the "radical interpretation" by some "that the complexity of the device and its mechanical sophistication put it so far beyond the scope of Hellenistic technology that it could only have been designed and created by alien astronauts coming from outer space and visiting our civilization."

Yet the fact is that nothing coming even close to the instrument's intricacy and precision has been found anywhere in any of the centuries preceding or following the time of the shipwreck. Even medieval astrolabes, more than a millennium after the Antikythera time frame, look like toys (Fig. 142a) compared to the ancient object (Fig. 142b). Moreover, the medieval and later European astrolabes and kindred devices were made of brass, which is easily malleable, whereas the ancient device was made of bronze—a metal useful in casting but extremely difficult to hone and shape in general and especially to produce a mechanism that is more intricate than modern chronometers.

Yet the instrument was there; and no matter who provided the science and technology for it, it proves that time-keeping and celestially guided navigation were possible at that early time at an incredible level of sophistication.

It seems that the reluctance to acknowledge the unacceptable also lies behind the fact that hardly anything con-

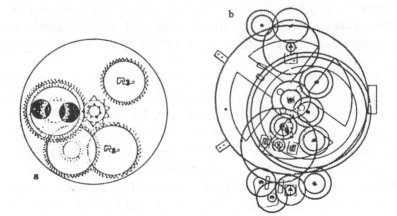

Figures 142a and 142b

cerning early cartography was brought up in the First Americans debate—even with such an opportunity as the 500th anniversary of the Columbus voyage in 1492.

Just across the Aegean Sea from Athens and the Kythera islands, in Istanbul (the previous Ottoman capital and the Byzantine one), in a converted palace now known as the Topkapi Museum, there is kept another find that throws light on ancient navigational capabilities. It is known as the *Piri Re'is Map,* after the Turkish admiral who had it made, and bears the Moslem year equivalent to A.D. 1513 (Fig. 143a). One of several *mapas mundi* (world maps) that have survived from that Age of Discovery, it attracted particular interest for a number of reasons: first, its accuracy and its sophisticated method of projecting global features on a flat surface; second, because it clearly shows (Fig. 143b) the whole of South America, with recognizable geographic and topographic features on both the Atlantic and Pacific coasts; and third, because it correctly projects the Antarctic continent. Although cartographed a few years after the Columbus voyages, the startling fact is that the southern parts of South America were unknown in 1513— Pizarro sailed from Panama to Peru only in 1530, and the Spaniards did not proceed farther down the coast or venture

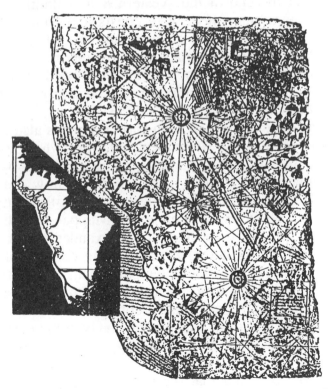

b

a

Figures 143a and 143b

inland to explore the Andean chain until years later. Yet the map shows all of South America, including its Patagonian tip. As to Antarctica, not only how it looks, but its very existence, was unknown until 1820—three *centuries* after the Piri Re'is map. Strenuous studies since the map was discovered in 1929 among the Sultan's treasures have reaffirmed these puzzling features of the map.

Brief notations on the map's margins are more fully explained in a treatise titled *Bahariyeh* ("About the Sea") that the admiral wrote. Regarding such geographic landmarks as the Antilles islands, he explained that he obtained the information from "the maps of the Genoese infidel, Colombo." He also repeated the tale of how Columbus first tried to convince the grandees of Genoa and then the king of Spain that according to a book that he (Columbus) possessed, "at the end of the Western Sea (Atlantic), that is on its western side, there were coasts and islands and all kinds of metals and also precious stones." This detail in the Turkish admiral's book confirms reports from other sources that Columbus knew quite well in advance where he was going, having come into possession of maps and geographic data from ancient sources.

In fact, the existence of such earlier maps is also attested to by Piri Re'is. In a subsequent notation, which explains how the map was drawn, he listed maps made by Arab cartographers, Portuguese maps ("which show the countries of Hind, Sind and China"), the "map of Columbus," as well as "about twenty charts and Mappae Mundi; these are charts drawn in the days of Alexander, Lord of the Two Horns." The latter was an Arabic epithet for Alexander the Great, and the statement means that Piri Re'is saw and used maps from the fourth century B.C. Scholars surmise that such maps were kept in the Library of Alexandria and that some must have survived the destruction by fire of that great hall of science by Arab invaders in A.D. 642.

It is now believed that the suggestion to sail westward on the Atlantic to reach existing coasts was first made not by Columbus but by an astronomer, mathematician, and

geographer from Florence, Italy, named Paulo del Pozzo
Toscanelli in 1474. It is also recognized that maps, such
as the Medicean from 1351 and that of Pizingi of 1367,
were available to later mariners and cartographers; the
most renowned of the latter has been Gerhard Kremer,
alias Mercator, whose *Atlas* of 1569 and methods of pro-
jection have remained standard features of cartography to
this day.

One of the odd things about Mercator's maps of the world
is that they show Antarctica, although that ice covered continent was not discovered, by British and Russian sailors,
until 250 years later, in 1820!

As those who had preceded (and succeeded) him, Mercator used for his *Atlas* earlier maps drawn by former
cartographers. In respect to the Old World, especially the
lands bordering on the Mediterranean, he obviously relied
on maps that went back to the time when Phoenicians and
Carthaginians ruled the seas, maps drawn by Marinus of
Tyre that were made known to future generations by the
astronomer, mathematician, and geographer Claudius Ptolemy who lived in Egypt in the second century A.D. For
his information on the New World, Mercator relied both
on olden maps and the reports of explorers since the
discovery of America. But where did he get the data not
only on the shape of Antarctica, but on its very existence?

Scholars agree that his probable source was a Map of the
World made in 1531 by Orontius Finaeus (Fig. 144a). Correctly projecting the Earth's globe by dividing it into the
northern and southern hemispheres, with the north and south
poles as epicenters, the map not only shows Antarctica—
an amazing fact by itself. It also shows Antarctica with
geographical and topographical features that have been buried under and obscured by an ice sheet for thousands of
years!

The map shows in unmistakable detail coasts, bays, in-
lets, estuaries and mountains, even rivers, where none are
now seen because of the ice cap that hides them. Nowadays
we know that such features exist, because they were discovered by scientific below-ice probings that culminated

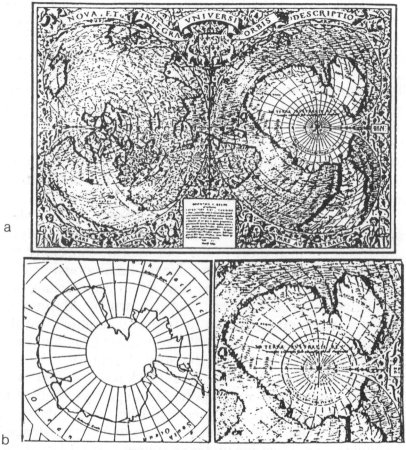

Figures 144a and 144b

with intensive surveys by many teams during the International Geophysical Year, 1958. The depiction on the Finaeus map, it then became clear, uncannily resembles the true shape of the Antarctic continent and its various geographical features (Fig. 144b).

In one of the most thorough studies of the subject, Charles H. Hapgood (*Maps of the Ancient Sea Kings*) concluded that the Finaeus map was drawn by him based on ancient charts that depicted Antarctica at a time when the continent, after having been freed of its ice covering, began to be covered by ice again in its western parts. That, his research team concluded, was about six thousand years ago, circa 4000 B.C.

Subsequent studies, as that by John W. Weihaupt (*Eos, the Proceedings of the American Geophysical Union*, August 1984), corroborated the earlier findings. Recognizing that "even crude mapping of a large continent would require a knowledge of navigation and geometry presumably beyond the ken of primitive navigators," he was nevertheless convinced that the map was based on data obtained some time between 2,600 and 9,000 years ago. The source of such data, he stated, remains an unanswered puzzle.

Presenting his conclusions in *Maps of the Ancient Sea Kings*, Charles Hapgood wrote: "It becomes clear that ancient voyagers traveled from pole to pole. Unbelievable as it may appear, the evidence nevertheless indicates that some ancient people explored Antarctica when its coasts were free of ice. It is clear, too, that they had an instrument of navigation for accurately determining longitudes that was far superior to anything possessed by the peoples of ancient, medieval, or modern times until the second half of the 18th century."

But those ancient mariners, as we have shown, only followed in the footsteps of the gods.

11

EXILES ON A
SHIFTING EARTH

Historians believe that exile as a deliberate penal policy was
introduced by the Assyrians in the eighth century B.C. when
they "carried off" kings, groups of elders and court officials, or even whole populations from their own lands to
live out their lives among strangers in far-off places. In fact,
the forced departure of someone into exile was a form of
punishment begun by the gods, and the first exiles were
leaders of the Anunnaki themselves. Such forced deportations, first of gods and then of people, have changed the
course of history. They also left their mark on the calendar,
and were linked to the coming of a New Age.

When the Spaniards, and then other Europeans, realized
how numerous were the similarities in traditions, customs,
and beliefs between those of the American natives and those
that have been associated with the Bible and the Hebrews,
they could think of no other explanation but that the "Indians" were descendants of the Ten Lost Tribes of Israel.
That harkened back to the mystery surrounding the whereabouts of the people belonging to the ten Israelite tribes that
formed the Northern Kingdom who were forced into exile
by the Assyrian king Shalmaneser. Biblical and postbiblical
sources held that, though dispersed, the exiles kept their
faith and customs so that they be counted among those who
will be redeemed and returned to their homeland. Ever since
the Middle Ages, travelers and savants claimed to have
found traces of the Ten Lost Tribes as far away as China
or as nearby as Ireland and Scotland. In the sixteenth century, the Spaniards were certain that it was such exiles who
had brought civilization to the Americas.

While the exile of the ten tribes by the Assyrians in the eighth century B.C., and then of the remaining two tribes by the Babylonians two centuries later, are historical facts, the "Ten Tribes Connection" to the New World remains in the realm of intriguing legends. Yet, unknowingly, the Spaniards were right in attributing the beginning of a formal civilization, with its own calendar, in the Americas to an exile; but not of a people—rather, the exile of a god.

The peoples of Mesoamerica—the Maya and Aztecs, Toltecs and Olmecs and lesser known tribes—had three calendars. Two were cyclical, measuring the cycles of the Sun and the Moon and of Venus. The other was chronological, measuring the passage of time from a certain starting point, "Point Zero." Scholars have established that this Long Count calendar's starting point was in the year that is designated under the Western calendar as 3113 B.C., but they know not what that starting point signifies. In *The Lost Realms* we have suggested that it marked the date of the arrival of Thoth, with a small band of aides and followers, in America.

Quetzalcoatl, the Great God of the Mesoamericans, was none other than Thoth, we have suggested. His epithet, the Plumed or Winged Serpent, was well known in Egyptian iconography (Fig. 145). Quetzalcoatl, like Thoth, was the god who knew and taught the secrets of temple building, numbers, astronomy, and the calendar. Indeed, the two other calendars of Mesoamerica by themselves offer clues for the Egyptian Connection and for identifying Quetzalcoatl

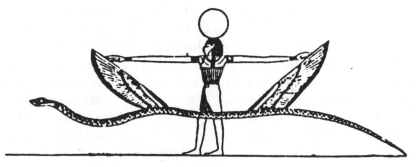

Figure 145

as Thoth. The two without doubt reveal the handiwork of "someone" familiar with the much earlier calendars of the Near East.

First of the two was the *Haab*, a solar-year calendar of 365 days that was subdivided into 18 months of 20 days each, plus an additional five special days at year's end. Although the 18 × 20 division is different from the Near Eastern one of 12 × 30, this calendar was basically an adaptation of the Egyptian calendar of 360 days plus 5. That purely solar calendar, as we have seen, was the one favored by Ra/Marduk; changing its subdivision could have been a deliberate act by Thoth to make it differ from that of his rival.

That purely solar calendar did not allow for intercalation—a device that, in Mesopotamia, was expressed in the addition of a thirteenth month once in a given number of years. In Mesoamerica this number, 13, featured in the next calendar.

As in Egypt, which had both a secular (pure solar-year) calendar as well as a sacred one, so was the second Mesoamerican calendar that of the Sacred Year called *Tzolkin*. In it the division into 20 also played a role; but it was counted in a cycle that rotated 13 times—the number inserted into the *Haab* calendar. That 13 × 20 resulted in a total of only 260 days. What this number, 260, represented or how it had originated has engendered many theories but no certain solutions. What is significant, calendrically and historically, is that these two cyclical calendars were meshed together, as gear wheels lock their teeth together (see Fig. 9b), to create the grand Sacred Round of fifty-two solar years; for the combination of 13, 20 and 365 could not repeat itself except once in 18,980 days, which meant fifty-two years.

This grand cycle of fifty-two years was sacred to all the peoples of Mesoamerica, and they related to it events both past and future. It lay at the core of the events associated with the greatest Mesoamerican deity, Quetzalcoatl ("The Plumed Serpent"), who having come to those lands from across the eastern seas was forced by the God of War to go into exile, but vowed to return in the year "1 Reed" of the fifty-two-year Sacred Cycle. In the Christian calendar, the parallel years were A.D. 1363, 1415, 1467, and 1519; the latter was the very year when Hernando Cortés appeared on the Mexican shores, fair-skinned and bearded as Quetzalcoatl had been; and so it was that the landing was seen by the Aztecs as fulfillment of the prophecy of the Returning God.

The centrality of the number fifty-two, if nothing else, as a hallmark of religious and messianic Mesoamerican beliefs and expectations, pointed to a key similarity between Quetzalcoatl and his Sacred Calendar and Thoth's calendar of fifty-two. The Game of Fifty-two was Thoth's game, and the Tale of Satni earlier related clearly stated that ''fifty-two was the magical number of Thoth.'' We have already explained the significance, in terms of Thoth's feud with Ra/Marduk, of the Egyptian calendar of fifty-two weeks. The Mesoamerican ''fifty-two'' had ''Thoth'' stamped all over it.

Another hallmark of Thoth was the application of a circular design to edifices related to the calendrical observations of the heavens. The Mesopotamian ziggurats were squarish, with their corners aligned to the cardinal points. Near Eastern temples—Mesopotamian, Egyptian, Canaanite, even Israelite—were rectangular structures whose axis was oriented either to the equinoxes or the solstices (a plan still manifest in churches and temples of our days). Only in the unique edifice that Thoth helped build in Lagash was a circular shape adopted. Its only other Near Eastern emulation was at the temple dedicated to Hathor (i.e. Ninharsag) at Denderah; and at Stonehenge, near where the Old World faces the New World on the other side of the Atlantic Ocean.

In the New World, in the domain of Adad, the younger son of Enlil and chief deity of the Hittites, the usual rectangular shape and orientation of Mesopotamian temples predominated. The greatest and oldest of them with its certain astronomical and calendrical functions, the Kalasasaya at Tiahuanacu, was rectangular and built on an east–west axis, not unlike the temple of Solomon. Indeed, one must wonder whether when the Lord took the prophet Ezekiel flying to show him an actual temple that was to serve as the model for the design of the future temple of Jerusalem, he did not fly him to Tiahuanacu to view the Kalasasaya, as the biblical detailed architectural text and a comparison of Figs. 50 and 124 may well suggest. Another major southern Andes temple, a focal point of sacred pilgrimages, was the one dedicated to the Great Creator that stood atop a promontory looking out to the expanse of the endless Pacific (not far south from present-day Lima). It too was rectangular in shape.

Judging by the design of these structures, Thoth was not invited there to take a hand in their construction. But if, as we believe, he was the Divine Architect of the circular observatories, he was certainly present in the Sacred Valley. His hallmarks among the structures of the Megalithic Age were the Round Observatory atop the Sacsahuaman promontory, the semicircular Holy of Holies in Cuzco, the Torreon in Machu Picchu.

The actual domain of Quetzalcoatl/Thoth was Mesoamerica and Central America, the lands of the Nahuatl-speaking and Mayan tribes; but his influence extended southward into the northern parts of the South American continent. Petroglyphs found near Cajamarca in the north of Peru (Fig. 146) that depict the Sun, the Moon, five-pointed stars, and other celestial symbols, show repeatedly next to them the symbol of the serpent—the unmistakable emblem of Enki and his clan and specifically so of the deity known as the "Plumed Serpent." The petroglyphs also include depictions of astronomical viewing devices, one held by a person (priest?), as was customary in the ancient Near East, and the other with the curved horns, as were the viewing devices erected in Egypt at the temples of Min (see Fig. 61).

The site appears to have been where ancient routes in the gold lands of the Andes, from the Pacific coast and from the Atlantic coast, met as the latter followed the rivers. Cajamarca itself, somewhat inland, and the natural harbor for it at Trujillo on the Pacific coast, in fact played a historic role in the European conquest of Peru. It was there, at

Figure 146

Trujillo, that Francisco Pizarro and his small band of soldiers landed in 1530. They marched inland and established their base at Cajamarca, a city "whose plaza was larger than any in Spain" and "whose buildings were three times the height of a man," according to the reports of the Conquistadors. It was to Cajamarca that the last Inca emperor, Atahualpa, was lured, only to be imprisoned for a ransom of gold and silver. The ransom was the filling up of a room twenty-five feet long and fifteen feet wide, as high as a man could not reach, with these precious metals. The ministers and priests of the king's retinue ordered that objects and artifacts made of gold and silver be brought from all over the land; S.K. Lothrop (*Inca Treasure as Depicted by Spanish Historians*) figured out that what the Spaniards then sent back to Spain from that ransom amounted to 180,000 ounces of gold and twice as much in silver. (Having collected the ransom, the Spaniards executed Atahualpa all the same.)

Farther north in Colombia, closer to Mesoamerica, at a site on the Magdalena River's banks, the petroglyphs unmistakably record Hittite and Egyptian encounters by including (Fig. 147) Hittite hieroglyphs (such as the "god" and "king" signs) alongside a variety of Egyptian symbols: cartouches (long rounded frames used to inscribe royal names), the hieroglyph for "splendor" (a circle with a dot in the center as the Sun with golden rays coming down), the "double Moon" Axe of Min.

Moving on northward, the Egyptian symbol par excellence, the drawing of a pyramid, is found among "graffiti" in the tomb area of Holmul, Guatemala (Fig. 148), thereby identifying the early inhabitants of Central America as familiar with Egypt. Also depicted is a circular stage-tower and next to it, apparently, its ground plan. It has all the appearance of a round observatory, similar to the one that had existed on the Sacsahuaman promontory down south.

Incredible as it may sound, reference to petroglyphs with astronomical symbols is made in ancient Near Eastern writings. The *Book of Jubilees*, enlarging and fleshing out the concise biblical record of the generations that followed the Deluge, describes how Noah instructed his descendants by relating to them the tale of Enoch and the knowledge that was granted him. The narrative continues thus:

Figure 147

Figure 148

In the twenty-ninth jubilee, in the first week, in the beginning thereof, Arpachshad took to himself a wife and her name was Rasu'eja, the daughter of Shushan, the daughter of Elam, and she bare him a son in the third year in this week, and he called his name Kainam.

And the son grew, and his father taught him writing, and he went to seek for himself a place which he might occupy as a city for himself.

And he found a writing which former generations had carved on the rock, and he read what there was thereon, and he transcribed it and sinned as a result thereof; for it contained the teaching of the Guardians in accordance with which they used to observe the omens of the sun and moon and stars in all the signs of heaven.

The petroglyphs, we learn from this millennia-old text, were not mere graffiti; they were expressions of knowledge of the "teachings of the Guardians"—the Anunnaki—"in accordance with which they used to observe the omens of the Sun and the Moon and the stars"; the petroglyphs were the "signs of heaven" of "former generations."

The depictions on rocks that we have just shown, including, as they do, round observatories, must be taken as eyewitness reports of what had actually been known and seen in antiquity in the Americas.

Indeed, in the heartland of Quetzalcoatl's domain, in Mexico, where the petroglyphs evolved into hieroglyphs akin to the earliest ones in Egypt, the most obvious traces of his presence are astronomically aligned temples, including circular and semicircular ones, and round observatories. Such remains begin with two perfectly round mounds that marked out the astronomical sightline at La Venta, one of the earliest sites of the Olmecs—African followers of Thoth who had arrived in Mexico by crossing the Atlantic circa 2500 B.C. At the other extreme of the four millennia that passed from then till the Spanish conquest, the latest instance of observatory-in-the-round was the semicircular pyramid in the sacred precinct of the Aztecs in Tenochtitlan (later Mexico City). It was so positioned that it served to determine the Day of the Equinox by watching, from the round "Tower of Quetzalcoatl," the Sun rise precisely between the opposite Pyramid of the Two Temples (Fig. 149).

Chronologically, between the early Olmecs and the late Aztecs, were the countless pyramids and sacred observatories of the Mayas. Some of them, as the one at Cuicuilco (Fig. 150a), were perfectly round. Others, like the one at

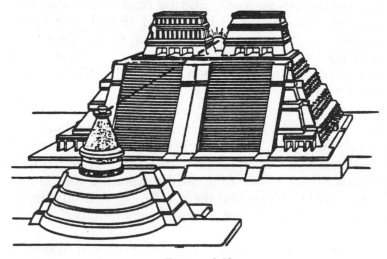

Figure 149

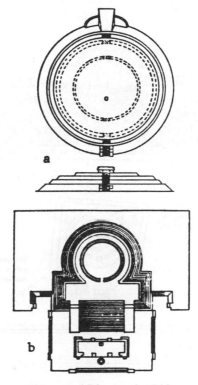

Figures 150a and 150b

Cempoala (Fig. 150b) began, as archaeologists have established, as purely round structures but in time changed shape as the small original stairways leading to their top stages evolved into monumental stairways and plazas. The most renowned of these structures is the *Caracol* in Chichén Itzá in the Yucatan peninsula—a circular observatory (Fig. 151) whose astronomical functions and orientations have been studied extensively and firmly established. Although the presently seen structure is believed to have been built only in A.D. 800 or thereabouts, it is known that the Mayas took over Chichén Itzá from earlier settlers, erecting Mayan structures where older ones used to be. An original observatory, scholars surmise, must have stood at the site at much earlier times, built-over and rebuilt as was the Mayan custom regarding the pyramids.

The sightlines in the existing structure have been exten-

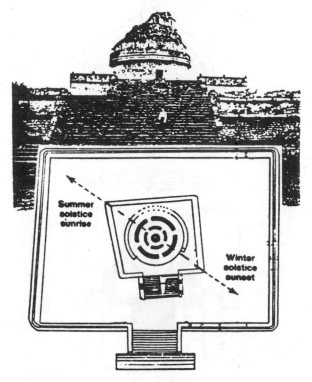

Figure 151

sively researched, and undoubtedly include the principal points of the Sun—the equinoxes and solstices, as well as some of the Moon's major points. Alignments with various stars are also suggested, though not with Venus; that is odd, for in Maya codices the movements of Venus are the principal subject. It is one of the reasons for believing that the sightlines were not devised by Mayan astronomers, but were inherited by them from previous eras.

The Caracol's ground plan—a round tower within a squarish enclosure as part of a larger rectangular structural frame, and the openings for sightlines in the tower itself—bring to mind the shape and layout (now seen only by their foundations) of the circular observatory within its square enclosure and larger rectangular complex at Sacsahuaman above Cuzco (see Fig. 120). Is there much doubt that both were designed by the same Divine Architect? In our opinion, he was Thoth.

For their observations Mayan astronomers used viewing devices that are often depicted in the codices (Fig. 152) and the similarities to Near Eastern instruments, viewing perches, and symbols are too numerous to be just coincidence. In all the instances the viewing perches are virtually identical to those atop Mesopotamian viewing towers or turrets; the symbol of the "stairway" that evolved from them, the ubiquitous symbol of the observatory at Tiahuanacu, is also clearly seen in the Mayan codices. One, in *Codex Bodley* (bottom of Fig. 152), indicates that the two astronomer-priests are viewing the Sun as it rises between two mountains; that is exactly how the Egyptian hieroglyphic texts depicted the idea and the word "horizon"; and it may not be by chance alone that the two mountains in the Mayan codex look like the two great pyramids of Giza.

The links with the ancient Near East in general, and with Egypt in particular, that are evidenced by glyphs and archaeological remains are augmented by legends.

The *Popol Vuh*, the "Council Book" of the highland Mayas, contains an account of how the sky and the Earth were formed, how the Earth was divided into four regions and partitioned, and how the measuring cord was brought

Figure 152

and stretched in the sky and over the Earth, creating the four corners. These are all elements basic to Near Eastern cosmogony and sciences, recollections of how the Earth was divided among the Anunnaki, and the functions of the Divine Measurers. Both Nahuatlan traditions, as well as Mayan ones in such forms as the Legend of Votan, recount the arrival of "the Fathers and the Mothers," the tribal ancestors, from across the seas. *The Annals of Cakchiquels,* a Nahuatlan record, states that while they themselves came from the west, there were others who had come from the east, in both instances "from the other side of the sea." The Legend of Votan, who had built the first city that was the cradle of Mesoamerican civilization, was written down by Spanish chroniclers from oral Mayan traditions. The emblem of Votan, they recorded, was the serpent; "he was a descendant of the Guardians, of the race of Can." "Guardians" was the meaning of the Egyptian term *Neteru* (i.e. "gods"). *Can,* studies such as that by Zelia Nuttal (*Papers of the Peabody Museum*) have suggested, was a variant of Canaan who was (according to the Bible) a member of the Hamitic peoples of Africa and a brother-nation of the Egyptians.

The possibility, which we have already mentioned, that the earliest migrants might have been descendants of *Cain,* relates Nahuatlan beginnings to one of the first recorded forced deportations: the exiling of Cain as punishment for the killing of Abel. The very first one, according to the Bible, was the expulsion of Adam and Eve from the Garden of Eden. In our times the exiling of kings has been a known occurrence; the exiling of Napoleon to the island of St. Helena is a notorious example. The biblical record shows that this mode of punishment goes back to the very beginnings of time, when Mankind was held to a certain code of ethics by the "gods." According to the earlier and more detailed Sumerian writings, it was in fact the gods themselves who applied the punishment to their own sinners; and the very first recorded instance concerned their commander in chief, Enlil: he was exiled to a land of banishment for the crime of raping a young Anunnaki nurse (in the end he married her and was given a reprieve).

It is clear from Nahuatlan and Mayan legends that Quetzalcoatl (Kukulkan in Mayan lore) had come to their lands with a small band of followers, and that his eventual departure was a forced one—an exile imposed by the War God. We believe that his arrival was also the result of a forced departure, an exile, from his native land, Egypt. The date of that first event is a vital component of the Mesoamerican counts of Time.

We have already discussed the centrality of the Sacred Round of fifty-two years in Mesoamerican calendrical, religious, and historical affairs, and shown that it was the sacred number of Thoth. Next in significance was a Grand Cycle of "perfect years" that encompassed thirteen eras of *baktuns*, units of four hundred years that were a key element in the consecutive calendar known as the Long Count.

The smallest unit in the Long Count calendar was the *kin*, a single day, and it was built up to larger numbers that could run into the millions of days by a series of multiplications, by 20 and by 360:

$$
\begin{array}{lll}
1\ \text{kin} & = & 1\ \text{day} \\
1\ \text{uinal} = 1\ \text{kin} \times 20 & = & 20\ \text{days} \\
1\ \text{tun} = 1\ \text{kin} \times 360 & = & 360\ \text{days} \\
1\ \text{ka-tun} = 1\ \text{tun} \times 20 & = & 7{,}200\ \text{days} \\
1\ \text{bak-tun} = 1\ \text{ka-tun} \times 20 & = & 144{,}000\ \text{days}
\end{array}
$$

As a purely arithmetical exercise the multiplications by twenty could continue, increasing the number of days that each term and its specific glyph represented, going on to 2,880,000 and 57,600,000 and so on. But in practice the Maya did not go beyond the *baktun;* for the count that began from the enigmatic starting point in 3113 B.C. was deemed to run in cycles of thirteen baktuns. Modern scholars divide the number of days that the Long Count indicates on Mayan monuments not by the perfect 360 but rather by the actual number 365.25 days of the solar year; thus, a monument stating "1,243,615" days is read to mean the passage of 3,404.8 years from August 3113 B.C., i.e. A.D. 292.

The concept of Ages in Earth's history and prehistory was a basic tenet of the pre-Columbian civilizations of Mesoamerica. According to the Aztecs, their Age or "Sun" was the fifth one and "began 5,042 years ago." While the Nahuatlan sources were not specific about how much longer this age was to last, the Mayan sources provided a more precise answer through the Long Count. The present "Sun," they said, will last precisely thirteen baktuns— 1,872,000 days from Point Zero. This represents a Grand Cycle of 5,200 "perfect years" of 360 days each.

In *The Mayan Factor* José Argüelles concluded that each baktun date had acted as a milestone in the history and prehistory of Mesoamerica, as will the year A.D. 2012, in which the thirteen baktuns that began in 3113 B.C. will be completed. He deemed the number 5,200 a key to understanding Mayan cosmogony and ages past and future.

In the 1930s Fritz Buck (*El Calendario Maya en la Cultura de Tiahuanacu*), seeing comparable elements between the Mayan calendars and that of Tiahuanacu, considered the starting date and other periodical markers to be related to actual events affecting the American peoples. He believed that a key symbol on the Gate of the Sun represented 52 and another one 520, and accepted as historically significant the number of 5,200 years; he held, however, that not one but two Great Cycles have to be considered, and that since 1,040 years remain in the second Great Cycle, the first one began in 9360 B.C. It was then, he held, that the legendary events and the tales of the gods in the Andes began. The second Great Cycle, accordingly, began at Tiahuanacu in 4160 B.C.

In arriving at A.D. 2012 as the end of the Fifth Sun, José Argüelles followed the present custom of dividing the 1,872,000 days by the actual number of 365.25 days in a solar year, resulting in the passage of only 5,125 years since the starting point in 3113 B.C. Fritz Buck on the other hand saw no need for such an adjustment, believing that the division should follow the Mayan 360 "perfect year." According to Buck, the historic age through which the Aztecs and Mayas had lived was to last a perfect 5,200 years.

This number, like fifty-two, is connected with Thoth according to ancient Egyptian sources. Among them were the writings of an Egyptian priest whom the Greeks called Manetho (his hieroglyphic name meant "Gift of Thoth"). He recorded the division of monarchies into dynasties, including the divine and semidivine ones that preceded the pharaonic dynasties; he also provided lengths of reign for all of them.

Corroborating legends and tales of the gods from other sources, the list by Manetho asserts that the seven great gods—Ptah, Ra, Shu, Geb, Osiris, Seth, and Horus—reigned a total of 12,300 years. Then began a second divine dynasty, headed by Thoth; it lasted 1,570 years. It was followed by thirty demigods who reigned 3,650 years. A chaotic time followed, a period of 350 years during which Egypt was disunited and in disarray. After that a person called Mên established the first pharaonic dynasty. Scholars hold that this happened circa 3100 B.C.

We have held that the actual date was 3113 B.C., the starting point of the Mesoamerican Long Count. It was then, we believe, that Marduk/Ra, reclaiming lordship over Egypt, expelled Thoth and his followers from that land, forcing them into exile in another, distant, land. And if the preceding reign of Thoth himself (1,570 years) and of his appointed demigods (3,650 years) is tallied, the result is 5,220 years—a mere discrepancy of 20 years from the 5,200 perfect years that make up the Great Mayan Cycle of thirteen baktuns.

As with 52, so was 5,200 a "number of Thoth."

In the olden days, when the Anunnaki were the Lords, the banishment and exile of gods marked milestones in what we have named *The Earth Chronicles*. Much of that part of the tale concerns Marduk, alias Ra in Egypt; and the calendar—the count of Divine, Celestial, and Earthly Time—played a major role in those events.

The reign of Thoth and his dynasty of demigods, ending

circa 3450 B.C., was followed in Egypt, according to Manetho, by a chaotic period that lasted 350 years, in the
aftermath of which dynastic rule by pharaohs beholden to
Ra began. Segments of the 175th chapter of the *Book of the
Dead* (known as the Papyrus of Ani) record an angry exchange between a reappearing Ra and Thoth. "O Thoth,
what is it that has happened?" Ra demanded to know. The
gods, he said, "have made an uproar, they have taken to
quarreling, they have done evil deeds, they have created
rebellion." They must have belittled Ra/Marduk in the
course of their rebellion: "They have made the great into
small."

Ra, the Great God, pointed an accusing finger at Thoth;
the accusation directly concerned changes in the calendar:
Thoth, Ra accused, "their years cut short, their months had
curbed." This Thoth had achieved by "the destruction of
Hidden Things that were made for them."

While the nature of the Hidden Things whose destruction
shortened the year and the months remains unknown, the
outcome could have only meant a switch from the longer
solar year to the shorter lunar year—the "making of the
great into small." The text ends with Thoth's accepting a
sentence of exile and banishment: "I am departing to the
desert, the silent land." It is such a tough place, the text
explains, that "sexual pleasures are not enjoyed in it" . . .

Another little-understood hieroglyphic text, found in one
of Tutankhamen's shrines as well as in royal tombs in
Thebes, may have recorded the expulsion order by Ra/Mar-
duk and gave among the reasons the calendrical conflict
between the "Sun god" and the "Moon god" (Thoth). The
text, which scholars are certain originated at a much earlier
time, relates how Ra ordered that Thoth be summoned to
him. When Thoth came before Ra, Ra announced: "Behold
ye, I am here in the sky in my proper place." Proceeding
to berate Thoth and "those who perform deeds of rebellion
against me," Ra told Thoth: "Thou encompasseth the two
heavens with thy shining rays; that is, Thoth as the Moon
encompasses." And he told Thoth: "I shall therefore have
thee go all the way around, to the place *Hau-nebut*." Some

scholars title the text ''The assignment of functions to Thoth.'' In fact, it was the ''assignment'' of Thoth to an unidentified distant land because of his ''functions''—calendrical preferences—relating to the Moon.

The exiling of Thoth was treated in Mesoamerican timekeeping as Point Zero of the Long Count—according to accepted chronology, in the year 3113 B.C. It must have been an event whose repercussions were recalled far and wide, for it could not be a mere coincidence that according to Hindu traditions (that also divide Earth's history and prehistory into Ages) the present Age, the *Kaliyuga,* began on a day equivalent to midnight between February 17 and 18 in 3102 B.C. This date is uncannily close to the date of Point Zero of the Mesoamerican Long Count, and is, therefore, in some way connected to the exiling of Thoth.

But no sooner than Marduk/Ra forced Thoth to leave the African domains, did he himself become the victim of a similar fate: exile.

With Thoth gone and his brothers Nergal and Gibil remote from the center of Egyptian power, Ra/Marduk could have expected an undisturbed supremacy there. But a new rival had emerged on the scene. He was Dumuzi, the youngest son of Enki, and his domain was the grasslands south of Upper Egypt. Unexpectedly, he emerged as a pretender to the Lordship over Egypt; and as Marduk soon discovered, the ambitions were prompted by a love affair of which Marduk was most disapproving. Preceding by millennia the setting and principals in Shakespeare's *Romeo and Juliet,* Dumuzi's bride was none other than Inanna/Ishtar, a granddaughter of Enlil and one who had fought alongside her brother and uncles to defeat the Enki'ites in the Pyramid Wars.

With limitless ambition, Inanna saw in the espousement with Dumuzi a great role for herself—if only he were to cease being just the Herder (as his epithet was) and assume lordship over the great Egyptian nation: ''I had a vision of a great nation choosing Dumuzi as God of its country,'' she later confided, ''for I have made Dumuzi's name exalted, I gave him status.''

Opposing the bethrothal and enraged by such ambitions, Marduk sent his "sheriffs" to arrest Dumuzi. Somehow the arrest went wrong; and Dumuzi, trying to hide in his sheep-folds, was found dead.

Inanna raised "a most bitter cry" and sought vengeance. Marduk, fearing her wrath, hid inside the Great Pyramid, all the while asserting his innocence because the death of Dumuzi was unintended, accidental. Unrelenting, Inanna "ceased not striking" at the pyramid, "at its corners, even its multitude of stones." Marduk issued a warning that he would resort to the use of awesome weapons "whose out-burst is terrible." Fearing another terrible war, the Anunnaki convened the supreme court of the Seven Who Judge. It was decided that Marduk must be punished, but since he did not directly kill Dumuzi, he could not be sentenced to death. The verdict was therefore to bury Marduk alive in the Great Pyramid within which he took refuge, by her-metically sealing it with him inside.

Various texts, quoted by us at length in *The Wars of Gods and Men,* relate the ensuing events, the commutaion of Marduk's sentence, and the dramatic effort to cut through the massive pyramid, using its original architectural draw-ings, to reach Marduk in time. The step-by-step rescue is described in detail. So is the conclusion of the incident: Marduk was sentenced to exile, and in Egypt Ra became *Amen*—the Hidden One, a god no longer seen.

As for Inanna, robbed by Dumuzi's death of her dream of being the Lady of Egypt, she was given Erech to be her "cult center" and the domain of Aratta to become the third region of civilization—that of the Indus Valley—circa 2900 B.C.

Where was Thoth in the ensuing centuries, now that his exiler was himself in exile? Apparently roaming distant lands—guiding the erection of the first Stonehenge in the British Isles circa 2800 B.C., helping orient astronomically megalithic structures in the Andes. Where was Marduk dur-ing that period? We really do not know, but he must have been somewhere not too far away, for he was watching developments in the Near East and continuing his scheming to seize the supremacy on Earth—supremacy, he believed, wrongly denied to his father Enki.

In Mesopotamia Inanna, ruthless and cunning, maneuvered the kingship of Sumer into the hands of a gardener whom she had found to be a man to her liking. She named him *Sharru-kin,* "righteous ruler," known to us as Sargon I. With Inanna's help he expanded his domains and created a new capital for a greater Sumer hence to be known as Sumer and Akkad. But seeking legitimacy, he went to Babylon—Marduk's city—and there removed some of its hallowed soil to use for foundations in his new capital. That was the opportunity for Marduk to reassert himself. "On account of the sacrilege thus committed," Babylonian texts recorded, "the great lord Marduk became enraged" and destroyed Sargon and his people; and, of course, reinstated himself in Babylon. There he began to fortify the city and enhance its underground water system, making it impervious to attack.

As the ancient texts reveal, it all had to do with Celestial Time.

Alarmed by the prospect of yet another devastating War of the Gods, the Anunnaki met in council. The chief antagonist was Ninurta, the heir apparent of Enlil, whose birthright Marduk was directly challenging. They invited Nergal, a powerful brother of Marduk, to join them in seeking a peaceful solution to the looming conflict. Mixing compliments with persuasion, Nergal first calmed down Ninurta, then agreed to go to Babylon similarly to persuade Marduk to step back from an armed confrontation. The chain of events, with dramatic and in the end fateful turns and consequences, is described in detail in a text known as the *Erra Epos* (Erra having been an epithet of Nergal). It includes many of the verbal exchanges between the participants as though a stenographer were present; and indeed, the text (as its postscript attests) was dictated to a scribe after the events by one of the participating Anunnaki.

As the story unfolds, it becomes increasingly clear that what was happening on Earth had been related to the heavens—to the constellations of the zodiac. In retrospect, the

statements and positions taken by the contestants for the
supremacy on Earth—Marduk the son of Enki and Ninurta
the son of Enlil—lead to no other conclusion than that *the
issue was the coming of a New Age:* the impending change
from the zodiac house of the bull (Taurus) to the zodiac
house of the ram (Aries) as the one in which the spring
equinox, and thus the calendrical moment for the New Year,
would occur.

Listing all his attributes and heirlooms, Ninurta thus
asserted:

> *In Heaven I am a wild bull,*
> *On Earth I am a lion.*
> *In the land I am the lord,*
> *among the gods I am the fiercest.*
> *The hero of the Igigi I am,*
> *among the Anunnaki I am powerful.*

The statement asserts verbally what the depictions, such
as we have shown in Fig. 93, have illustrated pictorially:
the zodiacal time when the spring equinox began in the
House of the Bull (Taurus) and the summer solstice occurred
in the zodiac of the Lion (Leo) belonged to the Enlilites,
whose "cult animals" were the Bull and the Lion.

Carefully, choosing his words, Nergal formulated his answer to the assertive Ninurta. Yes, he said, all that is true.
But

> *On the mountaintop,*
> *in the bush-thicket,*
> *see you not the Ram?*

Its emergence, Nergal continued, is unavoidable:

> *In that grove,*
> *even the supermost time measurer,*
> *the bearer of the standards,*
> *the course cannot change.*
> *One can blow like a wind,*

roar like a storm, [yet]
on the rim of the Sun's orbit,
no matter what the struggle,
see that Ram.

In its relentless precessional retardation, while the zodiacal constellation of the Bull was still dominant, "on the rim of the Sun's orbit" one could already see the approaching Age of the Ram.

But while the change was unavoidable, the time for it had not yet come. "The other gods are afraid of battle," Nergal said in conclusion. It could all be explained to Marduk, he felt. "Let me go and summon the prince Marduk away from his dwelling," make him leave peacefully, Nergal suggested.

And so, with Ninurta's reluctant consent, Nergal set out on a fateful mission to Babylon. On the way he stopped at Erech, seeking an oracle from Anu at his temple, the E.ANNA. The message he carried to Marduk from "the king of the gods" was this: *the Time has not yet come.*

The Time in question, the conversation-debate between Nergal and Marduk makes clear, was the impending zodiacal change—the coming of a New Age. Marduk received his brother in the E.SAG.IL, the ziggurat-temple of Babylon; the meeting took place in a sacred chamber called SHU.AN.NA, "The Celestially Supreme Place," which evidently Marduk deemed the most suitable place for the discussion; for he was certain that his time had come, and even showed Nergal the instruments he used to prove it. (A Babylonian artist, depicting the encounter between the two brothers, showed Nergal with his identifying weapon, and a helmeted Marduk standing atop his ziggurat and holding in his hand a device—Fig. 153—that looks very much like the viewing instruments that were employed in Egypt at the temples of Min.)

Realizing what had happened, Nergal argued to the contrary. Your "precious instrument," he told Marduk, was imprecise, and that is what had caused him to interpret incorrectly "the glow of the heavenly stars as the light of

Figure 153

the ordained day.'' While in your sacred precinct you have concluded that ''on the crown of your lordship the light did shine,'' it was not so at the Eanna, where Nergal had stopped on his way. There, Nergal said, ''the face of E.HAL.AN.KI in the Eanna remains covered over.'' The term E.HAL.AN.KI literally means ''House of the circling of Heaven-Earth'' and, in our view, suggests the location of instruments for determining the Earth's precessional shift.

But Marduk saw the issue differently. Whose instruments were really incorrect? At the time of the Deluge, he said, the ''regulations of Heaven-Earth shifted out of their groove and the stations of the celestial gods, the stars of heaven, have changed and did not return to their [former] places.'' A major cause of the change, Marduk claimed, was the fact that ''the *Erkallum* quaked and its covering was diminished, and the measures could no longer be taken.''

This is a highly significant statement, whose scientific importance—as that of the full text of the *Erra Epos*—has been ignored by scholars. *Erkallum* used to be translated ''Lower World'' and more recently the term is left intact, untranslated, as a word whose precise meaning is undetermined. We suggest that it is a term that denotes the land at the bottom of the world—*Antarctica;* and that the ''covering" or more literally "hair-growth-over" is a reference to the ice cap that, Marduk claimed, was still diminished millennia after the Deluge.

When it was all over, Marduk continued, he sent emissaries to check the Lower World. He himself went to take a look. But the "covering," he said, "had become hundreds of miles of water upon the wide seas": the ice cap was still melted.

This is a statement that corroborates our assertion, in *The 12th Planet,* that the Deluge was an immense tidal wave caused by the slippage of the Antarctic ice covering into the adjoining ocean, some 13,000 years ago. The event was the cause, we held, of the abrupt end of the last ice age and the climatic change it had brought about. It also left the Antarctic continent bare of its ice covering, enabling the seeing—and, evidently, mapping—of that continent as its landmass and shorelines really are.

The implication of Marduk's statement that the "regulations of Heaven-Earth had shifted out of their groove" as a result of the melting of the immense ice cap and the redistribution of its weight as water all over the world's seas, bears further study. Did it imply a change in Earth's declination? A somewhat different retardation and thus a different precessional schedule? Perhaps a slowing down of the Earth's spin, or of its orbit around the Sun? The results of experiments simulating the Earth's motions and wobbles with and without an Antarctic ice mass could be most illuminating.

All that, Marduk said, was aggravated by the fate of instruments in the Abzu, the southeastern tip of Africa. We know from other texts that the Anunnaki had a scientific station there that monitored the situation before the Deluge and was thus able to alert them to the impending calamity. "After the regimen Heaven-Earth was undone," Marduk continued, he waited until the fountains dried up and the floodwaters receded. Then he "went back and looked and looked; it was very grievous." What he had discovered was that certain instruments "that to Anu's heaven could reach" were missing, gone. The terms used to describe them are believed by scholars to refer to unidentified crystals. "Where is the instrument for giving orders?" he asked angrily, and "the oracle stone of the gods that gives the sign for lordship . . . Where is the holy radiating stone?"

These pointed questions regarding the missing precision instruments, which used to be operated by the "divine chief craftsman of Anu-powers who carried the holy All-Knower-of-the-Day," sound more like accusations than inquiries. We have earlier referred to an Egyptian text in which Ra/Marduk accused Thoth of destroying "the Hidden Things" that were used for determining the Earth's motions and calendar; the rhetorical questions thrown at Nergal imply deliberate wrongdoing against Marduk. In such circumstances, Marduk indicated, was he not right to rely on his own instruments to determine when *his* Time—the Age of the Ram—had arrived?

Nergal's full response is unclear because where it begins several lines on the tablet have been damaged. It appears that based on his own vast African domains, he did know where some of the instruments (or their replacements) were. He thus suggested that Marduk go to the indicated sites in the Abzu and verify it all for himself. He was certain that thereupon Marduk would realize that his birthright was not at risk; what was being challenged was the timing of his ascendancy.

To put Marduk further at ease, Nergal promised that he would personally see to it that nothing would be disturbed in Babylon during Marduk's absence. And, as a final gesture of reassurance, he promised to make the celestial symbols of the Enlilite Age, "the bulls of Anu and Enlil, crouch at the gate of thy temple."

Such a symbolic act of obeisance, the bowing to Marduk of Enlil's Bull of Heaven at the entrance to Marduk's temple, persuaded Marduk to accept his brother's plea:

Marduk heard this.
The promise, given by Erra [Nergal] found his favor.
So did he step down from his seat,

and to the Land of Mines, an abode of the Anunnaki,
he set his direction.

Thus did the dispute regarding the correct timing of the zodiacal change lead to Marduk's second exile—temporary only, he believed.

But as fate would have it, the anticipated coming of a New Age was not to be a peaceful one.

12

THE AGE OF THE RAM

When the Age of the Ram finally arrived, it did not come as the dawn of a New Age. Rather, it was accompanied by darkness at noon—the darkness of a cloud of deadly radiation from the first-ever explosion of nuclear weapons on Earth. It came as the culmination of more than two centuries of upheavals and warfare that pitted god against god and nation against nation; and in its aftermath, the great Sumerian civilization that had lasted for nearly two millennia lay prostrate and desolate, its people decimated, its remnants dispersed in the world's first Diaspora. Marduk did indeed gain supremacy; but the New Order that ensued was one of new laws and customs, a new religion and beliefs; an era of regression in sciences, of astrology instead of astronomy—even of a new and lesser status for women.

Did it have to happen that way? Was the change so devastating and bitter just because it involved ambitious protagonists—because the Anunnaki, not men, had directed the course of events? Or was it all destined, preordained, and the force and influence—real or imagined—of the passage into a new zodiacal house so overwhelming that empires must topple, religions must change, laws and customs and social organization must be overturned?

Let us review the record of that first known changeover; perchance we may find full answers, for sure enlightening clues.

It was, by our calculations, circa 2295 B.C. that Marduk left Babylon, going first to the Land of Mines and then to regions unspecified by the Mesopotamian texts. He left on the understanding that the instruments and other "works of wonder" that he had put up in Babylon would remain un-

318

disturbed; but no sooner did Marduk leave, than Nergal/ Erra broke his promise. Out of mere curiosity, or perhaps with malice in mind, he entered the forbidden *Gigunu*, the mysterious chamber that Marduk had declared off limits. Once inside he caused the chamber's "brilliance" to be removed; thereupon, as Marduk had warned, "the day turned into darkness," and calamities started afflicting Babylon and its people.

Was the "brilliance" a radiating, nuclear-driven device? It is not clear what it was, except that the adverse effects began to spread throughout Mesopotamia. The other gods were angered by Nergal's deed; even his father Enki reprimanded him and ordered him back to his African domain, Kutha. Nergal heeded the order; but before leaving he smashed all that Marduk had set up, and left behind his warriors to make sure that Marduk's followers in Babylon would remain subdued.

The two departures, first of Marduk and then of Nergal, left the arena free for the descendants of Enlil. First to take advantage of the situation was Inanna (Ishtar); she chose a grandson of Sargon, Naram-Sin ("Sin's Favorite") to ascend the throne of Sumer and Akkad; and with him and his armies as her surrogates, she embarked on a series of conquests. Among her first targets was the great Landing Place in the Cedar Mountains, the immense platform of Baalbek in Lebanon. She then assaulted the lands along the Mediterranean coast, seizing Mission Control Center in Jerusalem and the crossing point on the land route from Mesopotamia to the Sinai, Jericho. Now the spaceport itself, in the Sinai peninsula, was under her control. But, unsatisfied, Inanna sought to fulfill her dream of dominating Egypt—a dream shattered by the death of Dumuzi. Guiding, urging, and arming Naram-Sin with her "awesome weapons," she brought about the invasion of Egypt.

The texts suggest that recognizing her as an avowed adversary of Marduk, Nergal gave her his actual or tacit assistance in that invasion. But the other leaders of the Anunnaki did not view it all with equanimity. Not only did she breach the Enlilite-Enki'ite regional boundaries, she also brought under her control the spaceport, that neutral sacred zone in the Fourth Region.

An Assembly of the Gods was convened in Nippur to deal with Inanna's excesses. As a result, an order for her arrest and trial was issued by Enlil. Hearing that, Inanna forsook her temple in Agade, Naram-Sin's capital, and escaped to hide with Nergal. From afar, she sent orders and oracles to Naram-Sin, encouraging him to continue the conquests and bloodshed. To counteract that, the other gods empowered Ninurta to bring over loyal troops from neighboring mountainous lands. A text titled *The Curse of Agade* describes those events and the vow of the Anunnaki to obliterate Agade. True to that vow, the city—once the pride of Sargon and the dynasty of Akkad—was never to be found again.

The relatively brief Era of Ishtar had come to an end; and to bring some measure of order and stability to Mesopotamia and its neighboring lands, Ninurta (under whom Kingship had started in Sumer) was again given command of the country. Before Agade was destroyed, Ninurta its "crownband of lordship, the tiara of kingship, the throne given to rulership, to his temple brought over." At that time his "cult center" was in Lagash, at its Girsu sacred precinct. From there, flying in his Divine Black Bird, Ninurta roamed the plain between the two rivers and the adjoining mountainlands, restoring irrigation and agriculture, returning order and tranquility. Setting personal examples by his unwavering fidelity to his spouse Bau (nicknamed *Gula,* "the Great") with whom he had portraits made (Fig. 154), and devoted to his mother Ninharsag, he proclaimed moral laws and codes of justice. To assist in these tasks he appointed human viceroys; circa 2160 B.C., Gudea was the chosen one.

Over in Egypt, in the aftermath of the exile of Marduk/Ra, Naram-Sin's invasion and the reprimand to Nergal, the country was in disarray. Egyptologists call the chaotic century, between about 2180 and 2040 B.C., the "First Intermediate Period" in Egyptian history. It was a time when the Old Kingdom that was centered in Memphis and He-

Figure 154

liopolis came under attack from Theban princes in the south. Political, religious, and calendrical issues were involved; underlying the human contest was the celestial confrontation between the Bull and the Ram.

From the very beginning of Egyptian dynastic rule and religion, the greatest celestial compliment to the great gods was to compare them to the Bull of Heaven. Its earthly symbol, the Sacred Bull *Apis* (Fig. 155a) was venerated at Heliopolis and Memphis. Some of the earliest pictographic inscriptions—so old that Sir Flinders Petrie (*Royal Tombs*) attributed them to the time of "dynasty zero"—showed this symbol of the Sacred Bull upon a Celestial Boat with a priest holding ritual objects in front of it (Fig. 155b). (The depictions on this archaic plaque and on another similar one also reported by Sir Flinders Petrie, also clearly show the Sphinx, indicating beyond doubt that the Sphinx had already existed many centuries before its supposed construction by the Pharaoh Khephren of the Fourth Dynasty.) As later in Crete for the Minotaur, a special labyrinth was built for the Apis Bull in Memphis. At Saqqara, effigies of bull-heads made of clay with natural horns were placed in recesses within the tomb of a Second Dynasty pharaoh; and it is known that Zoser, a Third Dynasty pharaoh, held special ceremonies in honor of the Bull of Heaven at his spacious pyramid compound in Saqqara. All that had taken place during the Old Kingdom, a period that came to an end circa 2180 B.C.

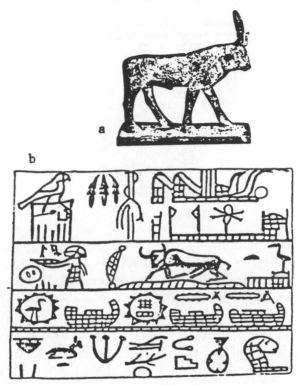

Figures 155a and 155b

When the Theban priests of Ra-Amen began the drive to supersede the Memphite-Heliopolitan religion and calendar, celestial depictions still showed the Sun rising over the Bull of Heaven (Fig. 156a), but the Bull of Heaven was depicted tethered and held back. Later on, when the New Kingdom reunited Egypt with Thebes as its capital and Amon-Ra was elevated to supremacy, the Bull of Heaven was depicted pierced and deflated (Fig. 156b). The Ram began to dominate celestial and monumental art and Ra was given the epithet "Ram of the Four Winds," and was so depicted to indicate that he was master of the four corners and four regions of the Earth (Fig. 157).

Where was Thoth during that First Intermediate Period, when in the heavens above and on Earth below the Ram and its followers were battling and chasing away the Bull and its adherents? There is no indication that he sought to

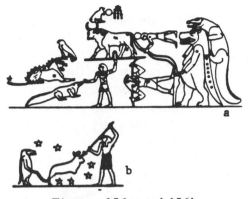

Figures 156a and 156b

Figure 157

reclaim the rulership of a divided and chaotic Egypt. It was a time when, without giving up his new domains in the New World, he could go about that in which he had become proficient—the erection of circular observatories and the teaching of the local inhabitants at old and new places the "secrets of numbers" and the knowledge of the calendar. The reconstruction of Stonehenge I into Stonehenge II and III at about that very time was one of those monumental edifices. If legends be deemed as conveyors of historical fact, then the one about Africans coming to erect the megalithic circles at Stonehenge suggests that Thoth, alias Quetzalcoatl, had brought over for the reconstruction task some of his Olmec followers who by then had become expert stonemasons in Mesoamerica.

The epitome of those undertakings was the invitation by Ninurta to come to Lagash and help design, orient, and build the Eninnu, Ninurta's new temple-pyramid.

Was it just a work of love, or was there a more compelling reason for that burst of astronomically related activity?

Dealing with the symbolism that guided Sumerian temple building, Beatrice Goff (*Symbols of Prehistoric Mesopotamia*) wrote thus of the construction of the Eninnu: "The time is the moment when in heaven and on earth the fates were decided." That the temple be built the way its divine planners had ordained and at the specific time it was to be built and inaugurated, she determined, was all "part of a plan foreordained when the fates were decided; Gudea's commission was part of a cosmic plan." This, she concluded, was "the kind of setting where not only art and ritual but also mythology go hand in hand as essentials in the religion."

Circa 2200 B.C. was indeed a time "when in Heaven and on Earth the fates were decided," for it was the time when a New Age, the Age of the Ram, was due to replace the Old Age, the Age of the Bull.

Though Marduk/Ra was somewhere in exile, there grew a contest for the hearts and minds of people since the "gods" had come to depend increasingly on human kings and human armies to achieve their ends. Many sources indicate that Marduk's son Nabu was crisscrossing the lands that later became known as Lands of the Bible, seeking adherents to his father's side. His name, *Nabu,* had the same meaning and came from the same verb by which the Bible called a true prophet: *Nabi,* one who receives the divine words and signs and in turn expresses them to the people. The divine signs of which Nabu spoke were the changing Heavens; the fact that the New Year and other worship dates no longer seemed to occur when they should have. Nabu's weapon, in behalf of Marduk, was the calendar . . .

What, one may ask, was there to view or determine that was unclear or in dispute? The truth of the matter is that even nowadays, no one can say for sure when one "Age" has ended and the other begun. There could be the arbitrary, mathematically precise calculation that since the Grand Precessional Cycle of 25,920 years is divided into twelve Houses, each House or Age lasts exactly 2,160 years. That was the mathematical basis of the sexagesimal system, the 10:6 ratio between Divine Time and Celestial Time. But if no person alive, no astronomer-priest, had witnessed the beginning of an Age and its ending, for no one human stayed alive 2,160 years, it was either the word of the gods, or the observation of the skies. But the zodiacal constellations are of varied sizes, and the Sun can linger longer or shorter periods within them. The problem is especially acute in the case of Aries, that occupies less than 30° of the celestial arc, while its neighbors Taurus and Pisces extend beyond their official 30° Houses. So, if the gods disagreed, some of them (e.g. Marduk, so well trained in sciences by his father Enki, and Nabu) could say: 2,160 years have passed, the Time has come. But others (e.g. Ninurta, Thoth) could and did say: But look to the Heavens, do you really see the change occurring?

The historical record, as detailed by the ancient texts and affirmed by archaeology, indicates that the tactics worked— at least for a while. Marduk remained in exile and in Mesopotamia the situation calmed down sufficiently for the mountainland troops to be sent back. After serving as a military headquarters for "ninety-one years and forty days" (according to the ancient records), Lagash could become a civilian center for the glorification of Ninurta. Circa 2160 B.C. that was expressed by the construction of the new Eninnu under Gudea's reign.

The Era of Ninurta lasted about a century and a half. Then, satisfied that the situation was under control, Ninurta departed for some distant mission. In his stead Enlil appointed his son Nannar/Sin to oversee Sumer and Akkad, and Ur, Nannar/Sin's "cult center," became the capital of a revitalized empire.

It was an appointment with more than political and hierarchical implications, for Nannar/Sin was the "Moon god" and his elevation to supremacy announced that the purely solar calendar of Ra/Marduk was done with and that the lunisolar calendar of Nippur was the only true one— religiously and politically. To assure adherence, a high priest knowledgeable in astronomy and celestial omens was sent from Nippur's temple to liaison at Ur. His name was Terah; with him was his ten-year-old son, Abram.

The year, by our calculations, was 2113 B.C.

The arrival of Terah and his family in Ur coincided with the establishment of the reign of five successive rulers known as the Ur III dynasty. Their, and Abram's, ensuing century saw on the one hand the glorious culmination of the Sumerian civilization; its epitome and hallmark was the grand ziggurat built there for Nannar/Sin—a monumental edifice that, though lying in ruins for almost four thousand years, still dominates the landscape and awes the viewer by its immensity, stability, and intricacy.

Under the active guidance of Nannar and his spouse Ningal, Sumer attained new heights in art and sciences, literature and urban organization, agriculture and industry and commerce. Sumer became the granary of the Lands of the Bible, its wool and garment industries were in a class by themselves, its merchants were the famed Merchants of Ur. But that was only one aspect of the Era of Nannar. On the other hand, hanging over all this greatness and glory was the destiny ordained by Time—the relentless change, from one New Year to another, of the Sun's position less and less in the House of GUD.ANNA, the "Bull of Heaven," and ever closer to that of KU.MAL, the celestial Ram— with all the dire consequences.

Ever since it was given Priesthood and Kingship, Mankind had known its place and role. The "gods" were the Lords, to be worshiped and venerated. There was a defined hierarchy, prescribed rituals, and holy days. The gods were strict but benevolent, their decrees were sharp but righteous. For millennia the gods oversaw the welfare and fate of Mankind, all the while remaining clearly apart from the people, approachable only by the high priest on specified dates, communicating with the king in visions and by omens. But now all that was beginning to crumble, for the gods themselves were at odds, citing different celestial omens and a changing calendar, increasingly pitting nation against nation in the cause of "divine" wars, quarrels, and bloodshed. And Mankind, confused and bewildered, increasingly speaking of "my god" and "your god," now even began to doubt the divine credibility.

In such circumstances Enlil and Nannar chose carefully the first ruler of the new dynasty. They selected Ur-Nammu ("The Joy of Ur"), a demigod whose mother was the goddess Ninsun. It was undoubtedly a very calculated move meant to evoke among the people memories of past glories and the "good old days," for Ninsun was the mother of the famed Gilgamesh who was still exalted in epic tales and artistic depictions. He was a king of Erech who was privileged to have seen both the Landing Place in the Cedar Mountains of Lebanon and the spaceport in the Sinai; and the choice of another son of Ninsun, some seven centuries later, was meant to evoke confidences that those vital places would again be part of Sumer's heritage, its Promised Lands.

Ur-Nammu's assignment was to steer the people "away from the evil ways" of following the wrong gods. The effort was marked by the repair and rebuilding of all the major temples in the land—with the conspicuous exception of Marduk's temple in Babylon. The next step was to subdue the "evil cities" where Nabu was making converts to Marduk. To that end Enlil provided Ur-Nammu with a "Divine Weapon" with which to "in the hostile lands heap up the rebels in piles." That the enforcement of the Enlilite Celestial Time was a major purpose is made clear in the text that quotes Enlil's instructions to Ur-Nammu about the weapon's use:

> *As the Bull*
> *to crush the foreign lands;*
> *As the Lion*
> *to hunt [the sinners] down;*
> *to destroy the evil cities,*
> *clear them of opposition to the Lofty Ones.*

The Bull of the equinox and the Lion of the solstice were to be upheld; any opponent of the Lofty Ones had to be hunted down, crushed, destroyed.

Leading the called-for military expedition, Ur-Nammu met not victory but an ignominious end. In the course of the battle his chariot got stuck in the mud and he fell off it, only to be crushed to death by its own wheels. The tragedy was compounded when the boat returning his body to Sumer sank on the way, so that the great king was not even brought to burial.

When the news reached Ur, the people were grieved and disbelieving. How did it happen that "the Lord Nannar did not hold Ur-Nammu by the hand," why did Inanna "not put her noble arm around his head," why did Utu not assist him? Why did Anu "alter his holy word"? Surely it was a betrayal by the great gods; it could only happen because "Enlil deceitfully changed his fate-decree."

The tragic death of Ur-Nammu and the doubting of the Enlilite gods at Ur caused Terah and his family to move to Harran, a city in northwestern Mesopotamia that served as a link with the lands and people of Anatolia—the Hittites; evidently, the powers that be felt that Harran, where a temple to Nannar/Sin almost duplicated that of Ur, would be a more appropriate place for the Nippurian scion of a priestly royal line in the turbulent times ahead.

In Ur, Shulgi, a son of Ur-Nammu by a priestess in a marriage arranged by Nannar, ascended the throne. He at once sought the favor of Ninurta, building for him a shrine in Nippur. The move had practical aspects; for as the western provinces became ever more restive in spite of a peace-journey undertaken by Shulgi, he arranged to obtain a "foreign legion" of troops from Elam, a Ninurta domain in the mountains southeast of Sumer. Using them to launch military expeditions against the "sinning cities," he himself sought solace in lavish living and lovemaking, becoming a "beloved" of Inanna and conducting banquets and orgies in Erech, in Anu's very temple.

Although the military expeditions brought, for the first time ever, Elamite troops to the gateway to the Sinai peninsula and its spaceport, they failed to quell the "rebellion" stirred up by Nabu and Marduk. In the forty-seventh year of his reign, 2049 B.Ċ., Shulgi resorted to a desperate stratagem: he ordered the building of a defensive wall along Sumer's western border. To the Enlilite gods it was tantamount to an abandonment of crucial lands where the Landing Place and Mission Control Center were. So, because "the divine regulations he did not carry out," Enlil decreed Shulgi's death, the "death of a sinner," the very next year.

The retreat from the western lands and the death of Shulgi triggered two moves. As we learn from a biographical text in which Marduk explained his moves and motives, it was then that he decided to return to the proximity of Mesopotamia by arriving in the land of the Hittites. Thereupon, it was also decided that Abram should make a move. In the forty-eight years of Shulgi's reign, Abram matured in Harran from a young bridegroom to a seventy-five-year-old leader, possessing varied knowledge and militarily trained and assisted by his Hittite hosts.

And Yahweh said unto Abram:
"Get thee out of thy country
and out of thy birthplace
and from thy father's house,
unto the land which I will show thee."
And Abram departed as Yahweh had spoken unto him.

The destination, as chapter 12 of Genesis makes clear, was the vital Land of Canaan; he was to proceed as quickly as possible and station himself and his elite cavalry in the Negev, on the Canaan-Sinai border. His mission, as we have fully detailed in *The Wars of Gods and Men,* was to protect the gateway to the spaceport. He arrived there skirting the "sinful cities" of the Canaanites; soon thereafter he went to Egypt, obtaining more troops and camels, for a cavalry, from the last pharaoh of the Memphite dynasties. Back in the Negev, he was ready to fulfill his mission of guarding the spaceport's approaches.

The anticipated conflict came to a head in the seventh

year of the reign of Shulgi's successor, Amar-Sin ("Seen by Sin"). It was, even in modern terms, a truly international war in which an alliance of four kings of the East set out from Mesopotamia to attack an alliance of five kings of Canaan. Leading the attack, according to the biblical record in chapter 14 of Genesis, was "Amraphel, the king of Shin'ar" and, for a long time, it was believed that he was the Babylonian king Hammurabi. In fact, as our own studies have shown, he was the Sumerian Amar-Sin and the tale of the international conflict has been recorded also in Mesopotamian texts, such as the tablets of the Spartoli Collection in the British Museum whose confirmation of the biblical tale was first pointed out by Theophilus Pinches in 1897. Together with complementary fragments, the collection of Mesopotamian tablets dealing with those events has come to be known as the *Khedorla' omer Texts*.

Marching under the banner of Sin and according to oracles given by Inanna/Ishtar, the allied army—probably the greatest military force of men ever seen until then—smote one western land after another. Regaining for Sin all the lands between the Euphrates and the Jordan River, they circled the Dead Sea and set as their next target the spaceport in the Sinai peninsula. But there Abram, carrying out his mission, stood in their way; so they turned back north, ready to attack the "evil cities" of the Canaanites.

Instead of waiting in their walled cities to be attacked, the Canaanite alliance marched forth and joined battle with the invaders in the Valley of Siddim. The records, both biblical and Mesopotamian, suggest an indecisive result. The "evil cities" were not obliterated, though the flight (and resulting death) of two kings, those of Sodom and Gomorrah, resulted in booty and prisoners being carried away from there. Among the prisoners from Sodom was Abram's nephew Lot; and when Abram heard that, his cavalry pursued the invaders, catching up with them near Damascus (now the capital of Syria). Lot, other prisoners, and the booty were retaken and brought back to Canaan.

As the Canaanite kings came out to greet them and Abram, they offered that he keep the booty as a reward.

But he refused to take "even a shoelace." He had acted neither out of enmity for the Mesopotamian alliance nor out of support for the Canaanite kings, he explained. It was only for "Yahweh, the God Most High, Possessor of Heaven and Earth, that I have raised my hand," he stated.

The unsuccessful military campaign depressed and confused Amar-Sin. According to the Date Formula for the ensuing year, 2040 B.C., he left Ur and the worship of Nannar/Sin and became a priest in Eridu, Enki's "cult center." Within another year he was dead, presumably of a scorpion's bite. The year 2040 B.C. was even more memorable in Egypt; there, Mentuhotep II, leader of the Theban princes, defeated the northern pharaohs and extended the rule and rules of Ra-Amen throughout Egypt, up to the Sinai boundary. The victory ushered in what scholars call the Middle Kingdom of the XI and XII dynasties that lasted to about 1790 B.C. While the full force and significance of the Age of the Ram came into play in Egypt during the later New Kingdom, the Theban victory of 2040 B.C. marked the end of the Age of the Bull in the African domains.

If, from a historical perspective, the coming of the Age of the Ram appears to have been inevitable, so must it also have appeared to the principal protagonists and antagonists of that very trying time. In Canaan, Abram retreated to a mountain stronghold near Hebron. In Sumer, the new king, Shu-Sin, a brother of Amar-Sin, strengthened the defensive walls in the west, sought an alliance with the Nippurites who had settled with Terah in Harran, and built two large ships—possibly as a precaution, with escape in mind . . . In a night equivalent to one in February 2031 B.C. a major lunar eclipse occurred in Sumer; it was taken to be an ominous omen of the nearing "eclipse" of the Moon god himself. The first victim, however, was Shu-Sin; for by the following year he was no longer king.

As the word of the celestial omen, the eclipse of the Moon, spread throughout the ancient Near East, the required messages of loyalty from viceroys and governors of the provinces, first in the west and then in the east, ceased.

Within a year of the reign of the next (and last) king of Ur, Ibbi-Sin, raiders from the west, organized by Nabu and encouraged by Marduk, were clashing with Elamite mercenaries at Mesopotamia's gates. In 2026 B.C. the compiling of customs receipts (on clay tablets) at Drehem, a major trade gateway in Sumer during the Ur III period, ceased abruptly, indicating that foreign commerce had come to a standstill. Sumer itself became a country under siege, its territory shrinking, its people huddled behind protective walls. In what was once the ancient world's food basket, supplies ran short and prices of essentials—barley, oil, wool—multiplied every month.

Unlike any other time in Sumer's and Mesopotamia's long history, omens were cited in unusual frequency. Judging by the record of human behavior one may see in that a known reaction to fear of the unknown and to a search for reassurance or guidance from some higher power or intelligence. But at that time there was a real cause for watching the heavens for omens, for the celestial arrival of the Ram was becoming increasingly evident.

As the texts that have survived from that period attest, the course of events about to happen on Earth was closely linked to celestial phenomena; and each side to the growing confrontation constantly observed the skies for heavenly signs. Since the various Great Anunnaki were associated with celestial counterparts, both zodiacal constellations and the twelve members of the Solar System (as well as with months), the movements and positions of the celestial bodies associated with the chief protagonists were especially significant. The Moon, counterpart of Ur's great god Nannar/Sin, the Sun (counterpart of Nannar's son Utu/Shamash), Venus (the planet of Sin's daughter Inanna/Ishtar), and the planets Saturn and Mars (associated with Ninurta and Nergal) were especially watched and observed in Ur and Nippur. In addition to all those associations, the various lands of the Sumerian empire were also deemed to belong, celestially, to specific zodiacal constellations: Sumer, Akkad, and Elam were under the sign and protection of Taurus; the Lands of the Westerners, under the sign of Aries. Hence, planetary and zodiacal conjunctions, sometimes coupled with the appearance (bright, dim, horned, etc.) of the Moon, Sun, and planets could spell good or evil omens.

A text designated by scholars *Prophecy Text B*, known from later copies of the original Sumerian record that was made in Nippur, illustrates how such celestial omens were interpreted as prophecies of the coming doom. In spite of breaks and damage, the impact of the tablet's text retains its predictions of the fateful events to come:

> *If [Mars] is very red, bright . . .*
> *Enlil will speak to the great Anu.*
> *The land [Sumer] will be plundered,*
> *The land of Akkad will . . .*
> *. . . in the entire country . . .*
> *A daughter will bar her door to her mother,*
> *. . . friend will slay friend . . .*
>
> *If Saturn will . . .*
> *Enlil will speak to the great Anu.*
> *Confusion will . . . troubles will . . .*
> *a man will betray another man,*
> *a woman will betray another woman . . .*
> *. . . a son of the king will . . .*
> *. . . temples will collapse . . .*
> *. . . a severe famine will occur . . .*

Some of those omen-prophecies directly related the planetary positions to the constellation of the Ram:

> *If the Ram by Jupiter will be entered*
> *when Venus enters the Moon,*
> *the watch will come to an end.*
> *Woes, troubles, confusion*
> *and bad things will occur in the lands.*
> *People will sell their children for money.*
> *The king of Elam will be surrounded in his palace:*
> *. . . the destruction of Elam and its people.*

If the Ram has a conjunction with the planet . . .
. . . when Venus . . . and the . . .
. . . planets can be seen . . .
. . . will rebel against the king,
. . . will seize the throne,
the whole land . . . will diminish at his command.

In the opposing camp, the heavens were also observed for signs and omens. One such text, put together through the labor of many scholars from assorted tablets (mostly in the British Museum), is an amazing autobiographical record by Marduk of his exile, agonizing wait for the right celestial omens, and final move to take over the Lordship that he believed was his. Written as a "memoir" by an aging Marduk, he reveals in it his "secrets" to posterity:

O great gods, learn my secrets
as I girdle my belt, my memories recall.
I am the divine Marduk, a great god.
I was cast off for my sins,
to the mountains I have gone.
In many lands I have been a wanderer;
From where the Sun rises to where it sets I went.

Having thus wandered from one end of the Earth to the other, he received an omen:

By an omen to Hatti-land I went.
In Hatti-land I asked for an oracle
[about] my throne and my Lordship.
In its midst [I asked]: "Until when?"
24 years in its midst I nested.

Various astronomical texts from the years that marked the transition from Taurus to Aries offer a clue regarding the omens that Marduk was especially interested in. In those texts, as well as in what is called by scholars "mythological texts," the association of Marduk with Jupiter is strongly suggested. We know that after Marduk had succeeded in his ambitions and established himself in Babylon as the supreme deity, such texts as the Epic of Creation were rewritten there so as to associate Marduk with Nibiru, the home planet of the Anunnaki. But prior to that Jupiter, by all indications, was the celestial body of Marduk in his epithet "Son of the Sun"; and a suggestion—made more than a century and a half ago—that Jupiter might have served in Babylon as a device parallel to that which Sirius had served in Egypt, as the synchronizer of the calendrical cycle, is quite pertinent here.

We refer to a series of lectures delivered at the Royal Institute of Great Britain to the Society of Antiquarians in 1822 (!) by an "antiquarian" named John Landseer in which, in spite of the meager archaeological data then available, he showed an astounding grasp of ancient times. Long before others, and as a result the holder of unaccepted views, he asserted that the "Chaldeans" had known of the phenomenon of precession millennia before the Greeks. Calling those early times an era "when Astronomy was Religion" and vice versa, he asserted that the calendar was related to the zodiacal "mansion" of the Bull, and that the transition to Aries was associated with "a mystifying conjunction of the Sun and Jupiter in the sign of Aries, at the commencement of the great cycle of intricate [celestial] revolutions." He believed that the Greek myths and legends connecting Zeus/Jupiter with the Ram and its golden fleece reflected that transition to Aries. And he calculated that such a determining conjunction of Jupiter and the Sun in the boundary between Taurus and Aries had occurred in the year 2142 B.C.

The notion that Jupiter in a conjunction with the Sun might have served as the Announcer, the herald of the Age of Aries, was also surmised from Babylonian astronomical tablets in a series of papers titled "Euphratean Stellar Researches" by Robert Brown in the *Proceedings of the Society of Biblical Archaeology*, London, in 1893. Focusing in particular on two astronomical tablets (British Museum catalogue numbers K.2310 and K.2894), Brown concluded that they dealt with the position of stars, constellations, and planets as seen in Babylon at midnight on a date equivalent to July 10, 2000 B.C. Apparently quoting Nabu in reference to his "proclamation of the planet of the Prince of Earth"—presumably Jupiter—appearing in an "ocular instance which took place in the sign of Aries," the texts were translated by Brown into a "star map" that showed Jupiter in near conjunction with the brightest star (*Lulim*, known by its Arabic name Hamal) of Aries and just off the point of the spring equinox, when the zodiacal path and the planetary path (celestial equator and ecliptic) cross (Fig. 158).

Dealing with the transitions from one Age to another as recorded in the Mesopotamian tablets, various Assyriologists (as they were called at the time)—e.g. Franz Xavier Kugler (*Im Bannkreis Babels*)—have pointed out that while the transition from Gemini to Taurus was ascertainable with relative precision, that from Taurus to Aries was less determinable timewise. Kugler believed that the vernal equinox signaling the New Year was still in Taurus in 2300 B.C., and noted that the Babylonians had assumed the *Zeitalter*, the new zodiacal Age, to have come into effect in 2151 B.C.

It is probably no coincidence that the same date marked an important innovation in Egyptian practices of depicting the heavens. According to the masterwork on the subject of ancient Egyptian astronomy, *Egyptian Astronomical Texts* by O. Neugebauer and Richard A. Parker, celestial

Fig. II.   Star-map in illustration of Tablet, K. 2310, Rev.

(Portion of the Midnight Sky as seen from Italytum, July 10, B.C. 2000.)

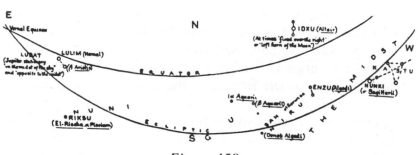

Figure 158

imaging including the thirty-six Decans began to be painted on coffin lids circa 2150 B.C.—coinciding with the chaotic First Intermediate Period, the start of the Theban push northward to supersede Memphis and Heliopolis, and the time when Marduk/Ra read the omens in his favor.

Coffin lids, as time went by and the Age of the Ram was no longer contested, clearly depicted the new Celestial Age, as this illustration from a tomb near Thebes shows (Fig. 159). The four-headed Ram dominates the four corners of the heavens (and the Earth too); the Bull of Heaven is shown pierced with a spear or lance; and the twelve zodiacal constellations, in their Sumerian-devised order and symbols, are arranged so that the constellation of Aries is precisely in the east, i.e., where the Sun appears on the Day of the Equinox.

If the determining or triggering omen for Marduk/Ra was

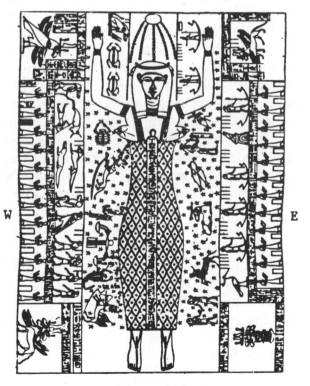

Figure 159

the conjunction of Jupiter and the Sun in the "mansion" of Aries, and if it did occur in 2142 B.C. as John Landseer suggested, then this heralding more or less coincided with the arithmetically calculated (once in 2,160 years) zodiacal shift. That, however, would have meant that the claim that the shift to Aries had come about preceded by about a century and a half the *observational* shift of the vernal equinox into Aries in 2000 B.C. as attested by the two tablets. That discrepancy could explain, at least in part, the disagreement at that time regarding what the celestial omens or observations were truly portending.

As the autobiographical Marduk text admits, even the omen that signified to him the time to end his wanderings and come to Hatti Land, the Land of the Hittites in Asia Minor, occurred twenty-four years before his next move. But that and other celestial omens were also watched closely on the Enlilite side; and although the Ram had not yet fully dominated the New Year's day on the spring equinox in the time of Ibbi-Sin, the last king of Ur, the oracle priests interpreted the omens as portents of the disastrous end. In the fourth year of Ibbi-Sin's reign (2026 B.C.) the oracle priests told him that according to the omens, "For the second time, he who calls himself Supreme, like one whose chest has been anointed, shall come from the west." With such predictions Sumerian cities, in the fifth year of Ibbi-Sin's reign, ceased the delivery of the traditional sacrificial animals for Nannar's temple in Ur. That same year the omen-priests prophesied that "when the sixth year comes, the inhabitants of Ur will be trapped." In the following, sixth year, the omens of destruction and ruin became more urgent and Mesopotamia itself, the heartland of Sumer and Akkad, was invaded. The inscriptions record that in the sixth year the "hostile Westerners had entered the plain, had entered the interior of the country, taking one by one all the great fortresses."

In the twenty-fourth year of his sojourn in the Land of the Hittites, Marduk received another omen: "My days [of exile] were completed, my years [of exile] were fulfilled," he wrote in his memoirs. "With longing to my city Babylon I set course, to my temple Esagila as a mount [to rebuild], my everlasting abode to reestablish." The partly damaged tablet then describes Marduk's route from Anatolia back to Babylon; the cities named indicate that he first went south to Hama (the biblical Hamat), then crossed the Euphrates at Mari; he indeed returned, as the omens had predicted, from the west.

The year was 2024 B.C.

In his autobiographical memoirs Marduk described how he had expected his return to Babylon to be a triumphant one, opening an era of well-being and prosperity for its people. He envisaged the establishment of a new royal dynasty, and foresaw as the first task of the new king the rebuilding of the Esagil, the temple-ziggurat of Babylon, according to a new "ground plan of Heaven and Earth"— one in accord with the New Age of the Ram:

> *I raised my heels toward Babylon,*
> *through the lands I went to my city;*
> *A king in Babylon to make the foremost,*
> *in its midst my temple-mountain to heaven raise.*
> *The mountainlike Esagil he will renew,*
> *the ground plan of Heaven and Earth*
> *will he for the mountainlike Esagil draw,*
> *its height he will alter,*
> *its platform he will raise,*
> *its head he will ameliorate.*
>
> *In my city Babylon*
> *in abundance he will reside;*
> *My hand he will grasp,*
> *to my city and my temple Esagil*
> *for eternity I shall enter.*

Undoubtedly mindful of the manner in which Ninurta's ziggurat-temple at Lagash was decorated and embellished, Marduk envisioned his own new temple, the Esagil ("House whose head is loftiest"), decorated with bright and precious metals: "with cast metal will it be covered, its steps with drawn metal will be overlaid, its sidewalls with brought-over metal will be filled." And when all that shall be completed, Marduk mused, astronomer-priests shall ascend the ziggurat's stages and observe the heavens, confirming his rightful supremacy:

> *Omen-knowers, put to service,*
> *shall then ascend its midst;*
> *Left and right, on opposite sides,*
> *they shall separately stand.*
> *The king will then approach;*
> *the rightful star of the Esagil*
> *over the land [he will observe].*

When the Esagil was eventually built, it was erected according to very detailed and precise plans; its orientation, height, and various stages were indeed such that its head pointed directly (see Fig. 33) to the star *Iku,* the lead star of the constellation Aries.

But Marduk's ambitious vision was not to be fulfilled right then and there. In the very same year that he began his march back to Babylon at the head of a horde of Western supporters organized by Nabu, a most awesome catastrophe befell the ancient Near East—a calamity the likes of which neither Mankind nor Earth itself had previously experienced.

He expected that once the omens were clear, both gods and men would heed his call for accepting his supremacy without further resistance. "I called on the gods, all of them, to heed me," Marduk wrote in his memoirs. "I called on the people along my march, 'bring your tribute to Babylon.'" Instead, he encountered a scorched-earth policy: the gods in charge of cattle and grains left, "to heaven they went up," and the god in charge of beer "made sick the heart of the land." The advance turned violent and bloody. "Brother consumed brother, friends slew each other with the sword, corpses of people blocked the gates." The land was laid waste, wild animals devoured people, packs of dogs bit people to death.

As Marduk's followers continued their advance, the temples and shrines of other gods began to be desecrated. The greatest sacrilege was the defilement of Enlil's temple in Nippur, until then the venerated religious center of all the lands and all the peoples. When Enlil heard that even the Holy of Holies was not spared, that "in the holy of holies the veil was torn away," he rushed back to Mesopotamia. He "set off a brilliance like lightning" as he came down from the skies; "riding in front of him were gods clothed with radiance." Seeing what had happened, "Enlil evil against Babylon caused to be planned." He ordered that Nabu be seized and brought before the Council of the Gods, and Ninurta and Nergal were given the assignment. But they found that Nabu had escaped from his temple in Borsippa, on the Euphratean border, to hide among his followers in Canaan and the Mediterranean islands.

Meeting in Council, the leading Anunnaki debated what to do, discussing the alternatives "a day and a night, without ceasing." Only Enki spoke up in defense of his son: "Now that prince Marduk has risen, now that the people for the second time have raised his image," why does opposition continue? He reprimanded Nergal for opposing his brother; but Nergal, "standing before him day and night without cease," argued that the celestial omens were being misread. "Let Shamash"—the Sun god—"see the signs and inform the people," he said; "Let Nannar"—the Moon god—"at his sign look and impart that to the land." Referring to a constellation-star whose identity is being debated, Nergal said that "among the stars of heaven the Fox Star was twinkling its rays to him." He was seeing other omens—"dazzling stars of heaven that carry a sword"—comets streaking in the skies. He wanted to know what these new omens meant.

As the exchanges between Enki and Nergal became harsher, Nergal, "leaving in a huff," announced that it was necessary to "activate that which with a mantle of radiance is covered," and thereby make the "evil people perish."

There was no way to block the takeover by Marduk and Nabu except by the use of "the seven awesome weapons," whose hiding place in Africa he alone knew. They were weapons that of the lands could make "a dust heap," cities "to upheaval," seas "to agitate, that which teems in them to decimate" and "people make vanish, their souls turn to vapor." The description of the weapons and the consequences of their use clearly identifies them as nuclear weapons.

It was Inanna who had pointed out that time was running out. "Until the time is fulfilled, the hour will be past!" she told the arguing gods; "pay attention, all of you," she said, advising them to continue their deliberations in private, lest the plan of attack be divulged to Marduk (presumably by Enki). "Cover your lips," she told Enlil and the others, "go into your private quarters!" In the privacy of the Emeslam temple, Ninurta spoke up. "The time has elapsed, the hour has passed," he said. "Open up a path and let me take the road!"

The die was cast.

Of the various extant sources dealing with the fateful chain of events, the principal and most intact one is the *Erra Epic*. It describes in great detail the discussions, the arguments for and against, the fears for the future if Marduk and his followers should control the spaceport and its auxiliary facilities. Details are added by the *Khedorlaomer Texts* and inscriptions on various tablets, such as those in the *Oxford Editions of Cuneiform Texts*. They all describe the ominous and fateful march to its culmination, of which we can read in Genesis, chapters 18 and 19: the "upheavaling" of Sodom and Gomorrah and of the "evil cities" of their plain, "and all the inhabitants of the cities, and all that which grew on the ground."

The upheavaling and wiping off the face of the Earth of the "evil cities" was only a sideshow. The main target of obliteration was the spaceport in the Sinai peninsula. "That which was raised toward Anu to launch," the Mesopotamian texts state, Ninurta and Nergal "caused to wither; its face they made fade away, its place they made desolate." The year was 2024 B.C.; the evidence—the immense cavity in the center of the Sinai and the resulting fracture lines, the vast surrounding flat area covered with blackened stones, traces of radiation south of the Dead Sea, the new extent and shape of the Dead Sea—is still there, four thousand years later.

The aftereffects were no less profound and lasting. The nuclear blasts and their brilliant flashes and earthshaking impact were neither seen nor felt far away in Mesopotamia; but as it turned out, the attempt to save Sumer, its gods, and its culture in fact led to a dismal end for Sumer and its civilization.

The bitter end of Sumer and her great urban centers is described in numerous Lamentation Texts, long poems that bewail the demise of Ur, Nippur, Uruk, Eridu, and other famed and less famed cities. Typical of the calamities that befell the once proud and prosperous land are those listed in the *Lamentation Over the Destruction of Ur,* a long poem of some 440 verses of which we shall quote but a few:

The city into ruins was made,
the people groan . . .
Its people, not potsherds,
filled its ravines . . .
In its lofty gates, where they were wont
to promenade, dead bodies lay about . . .
Where the festivities of the land took place,
the people lay in heaps . . .
The young were lying in their mothers' laps
like fish carried out of the waters . . .
The counsel of the land was dissipated.

In the storehouses that abounded in the land,
fires were kindled . . .
The ox in its stable has not been attended,
gone is its herdsman . . .
The sheep in its fold has not been attended,
gone is its shepherd boy . . .
In the rivers of the city dust has gathered,

into fox dens they have become . . .
In the city's fields there is no grain,
gone is the fieldworker . . .
The palm groves and vineyards, with honey and wine
abounded, now bring forth mountain thorns . . .
Precious metals and stones, lapis lazuli,
have been scattered about . . .
The temple of Ur has been given over
to the wind . . .
The song has been turned into weeping . . .
Ur has been given over to tears.

For a long time scholars have held the view that the
various lamentation texts dealt with the successive but separate destruction of Sumer's cities by invaders from the
west, the east, the north. But in *The Wars of Gods and Men*
we have suggested that it was not so; that what these lamentations deal with was one single countrywide calamity,
an unusual catastrophe and a sudden disaster against which
no protection, no defense, no hiding was possible. This
view, of a single sudden and overwhelming calamity, is
now increasingly accepted by scholars; yet to be accepted
is the evidence that we have presented that the calamity was
linked to the "upheavaling" of the "evil cities" and the
spaceport in the west. It was the unexpected development
of an atmospheric vacuum, creating an immense whirlwind
and a storm that carried the radioactive cloud eastward—
toward Sumer.

The various available texts, and not just the lamentation
texts, clearly speak of the calamity as an unstoppable storm,
an Evil Wind, and clearly identify it as the result of an
unforgettable day when a nuclear blast had created it near
the Mediterranean coast:

> *On that day,*
> *When heaven was crushed*
> *and the Earth was smitten,*
> *its face obliterated by the maelstrom—*

When the skies were darkened
and covered as with a shadow—

On that day there was created

A great storm from heaven . . .
A land-annihilating storm . . .
An evil wind, like a rushing torrent . . .
A battling storm joined by a scorching heat . . .
By day it deprived the land of the bright sun,
in the evening the stars did not shine . . .

The people, terrified, could hardly breathe;
The Evil Wind clutched them,
does not grant them another day . . .
Mouths were drenched in blood,
heads wallowed in blood . . .
The face was made pale by the Evil Wind.

After the deadly cloud had moved on, ''after the storm was carried off from the city, that city was turned into desolation'':

It caused cities to be desolated,
It caused houses to become desolate,
It caused stalls to become desolate,
the sheepfolds to be emptied . . .
Sumer's rivers it made flow
with water that is bitter;
its cultivated fields grow weeds,
its pastures grow withering plants.

It was a death-carrying storm that endangered even the gods. The lamentations list virtually every major Sumerian city as places where their gods had abandoned their abodes, temples, and shrines—in most cases never to return. Some escaped the approaching cloud of death hurriedly, ''flying off as a bird.'' Inanna, having rushed to sail off to a safe haven, later complained that she had to leave behind her

jewelry and other possessions. The story, however, was not
the same everywhere. In Ur, Nannar and Ningal refused to
abandon their followers and appealed to the great Enlil to
do whatever possible to avert the disaster, but Enlil re-
sponded that the fate of Ur could not be changed. The divine
couple spent a nightmarish night in Ur: "Of that night's
foulness they did not flee," hiding underground "as ter-
mites." But in the morning Ningal realized that Nannar/Sin
had been afflicted, and "hastily putting on a garment" departed the beloved Ur with the stricken mate. In Lagash,
where with Ninurta away Bau had stayed in the Girsu by
herself, the goddess could not force herself to leave. Lingering behind "she wept bitterly for her holy temple, for
her city." The delay almost cost her her life: "On that day,
the storm caught up with her, with the Lady." (Indeed,
some scholars deem the ensuing verse in the lamentation to
indicate that Bau had in fact lost her life: "Bau, as if she
were a mortal, the storm had caught up with her.")

Fanning out in a wide swath over what used to be Sumer
and Akkad, the Evil Wind's path touched Eridu, Enki's
city, in the south. Enki, we learn, took cover some distance
away from the wind's path, yet close enough to be able to
return to the city after the cloud had passed. He found a
city "smothered with silence, its residents stacked up in
heaps." But here and there there were survivors, and Enki
led them southward, to the desert. It was an "inimical
land," uninhabitable; but using his scientific prowess,
Enki—like Yahweh half a millennium later in the Sinai
desert—managed miraculously to provide water and food
for "those who have been displaced from Eridu."

As fate would have it, Babylon, situated on the northern
edge of the Evil Wind's wide swath, was the least affected
of all the Mesopotamian cities. Alerted and advised by his
father, Marduk urged the city's people to leave and hurry
northward; and, in words reminiscent of the angels' advice
to Lot and his family as they were told to leave Sodom
before its upheavaling, Marduk told the escapees "neither
to turn nor to look back." If escape was not possible, they
were told to "get thee into a chamber below the earth, into

a darkness." Once the Evil Storm had passed, they were not to consume any of the food or beverage in the city, for they might have been "touched by the ghost."

When the air finally cleared, all of southern Mesopotamia lay prostrate. "The storm crushed the land, wiped out everything . . . No one treads the highways, no one seeks out the roads . . . On the banks of the Tigris and the Euphrates, only sickly plants grow . . . In the orchards and the gardens there is no new growth, quickly they waste away . . . On the steppes cattle large and small become scarce . . . The sheepfolds have been delivered to the Wind."

Life began to stir anew only seven years later. Backed by Elamite and Gutian troops loyal to Ninurta, a semblance of organized society returned to Sumer under rulers seated in former provincial centers, Isin and Larsa. It was only after the passage of seventy years—the same interval that later applied to the restoration of the temple in Jerusalem—that the temple in Nippur was restored. But the "gods who determine the destinies," Anu and Enlil, saw no purpose in resurrecting the past. As Enlil had told Nannar/Sin who had appealed in behalf of Ur—

> *Ur was granted kingship—*
> *it was not granted an eternal reign.*

Marduk had won out. Within a few decades, his vision of a king in Babylon who would grasp his hand, rebuild the city, raise high its ziggurat Esagil—had come true. After a halting start, the First Dynasty of Babylon attained the intended power and assurance that were expressed by Hammurabi:

> *Lofty Anu, lord of the gods*
> *who from Heaven to Earth came,*
> *and Enlil, lord of Heaven and Earth*
> *who determines the destinies of the land,*
> *Determined for Marduk, the firstborn of Enki,*
> *the Enlil-functions over all mankind;*

Made him great among the gods who watch and see,
Called Babylon by name to be exalted,
made it supreme in the world;
And established for Marduk, in its midst,
an everlasting kingship.

In Egypt, unaffected by the nuclear cloud, the transition
to the Age of the Ram began right after the Theban victory
and the enthronement of the Middle Kingdom dynasties.
When the celebrations of the New Year, coinciding with
the rising of the Nile, were adjusted to the New Age, hymns
to Ra-Amen praised him thus:

O Brilliant One
who shines in the inundation waters.
He who raised his head and lifts his forehead:
He of the Ram, the greatest of celestial creatures.

Under the New Kingdom, temple avenues were lined with
statues of the Ram; and in the great temple to Amon-Ra in
Karnak, in a secret observation perch that had to be opened
on the day of the winter solstice to let in the Sun's rays
through the path to the Holy of Holies, the following instructions were inscribed for the astronomer-priest:

One goes toward the hall called Horizon of the Sky.
One climbs the Aha, *"Lonesome place of the majestic*
 soul,"
the high room for watching the Ram who sails across the
 skies.

In Mesopotamia, slowly but surely, the ascendancy of
the Age of the Ram was recognized by changes in the
calendar and in the lists of the celestial stars. Such lists,
that used to begin with Taurus, now began with Aries; and
for Nissan, the month of the spring equinox and the New
Year, the zodiac of Aries rather than Taurus was thereafter
written in. An example is the Babylonian astrolabe ("taker
of stars") that we have discussed earlier (see Fig. 102) in

Figure 160

connection with the origin of the division into thirty-six segments. It clearly inscribed the star *Iku* as the defining celestial body for the first month Nisannu. Iku was the "alpha" or lead star of the constellation of the Ram; it is still known by its Arabic name *Hamal,* meaning "male sheep."

The New Age had arrived, in the heavens and on Earth.

It was to dominate the next two millennia and the astronomy that the "Chaldeans" had transmitted to the Greeks. When, in the closing years of the fourth century B.C., Alexander came to believe that he was entitled—like Gilgamesh 2,500 years earlier—to immortality because his true father was the Egyptian god Amon, he went to the god's oracle place in Egypt's western desert to seek confirmation. Having received it, he struck silver coins bearing his image adorned with the horns of the Ram (Fig. 160).

A few centuries later the Ram faded and was replaced by the sign of the Fishes, Pisces. But that, as the saying goes, is already history.

13

AFTERMATH

To establish his supremacy on Earth, Marduk proceeded to establish his supremacy in the heavens. A major vehicle to that end was the all-important annual New Year celebration, when the Epic of Creation was read publicly. It was a tradition whose purpose was to acquaint the populace not only with the basic cosmogony and the tale of Evolution and the arrival of the Anunnaki, but also as a way to state and reinstate the basic religious tenets regarding Gods and Men.

The Epic of Creation was thus a useful and powerful vehicle for indoctrination and reindoctrination; and as one of his first acts Marduk instituted one of the greatest forgeries ever: the creation of a Babylonian version of the epic in which the name "Marduk" was substituted for the name "Nibiru." It was thus Marduk, as a celestial god, who had appeared from outer space, battled Tiamat, created the Hammered Out Bracelet (the Asteroid Belt) and Earth of Tiamat's halves, rearranged the Solar System, and became the Great God whose orbit encircles and embraces "as a loop" the orbits of all the other celestial gods (planets), making them subordinate to Marduk's majesty. All the ensuing celestial stations, orbits, cycles, and phenomena were thus the masterworks of Marduk: it was he who determined Divine Time by his orbit, Celestial Time by defining the constellations, and Earthly Time by giving Earth its orbital position and tilt. It was he, too, who had deprived Kingu, Tiamat's chief satellite, of its emerging independent orbit and made it a satellite of Earth, the Moon, to wax and wane and usher in the months.

In so rearranging the heavens, Marduk did not forget to settle some personal accounts. In the past Nibiru, as the home planet of the Anunnaki, was the abode of Anu and thus associated with him. Having appropriated Nibiru to himself, Marduk relegated Anu to a lesser planet—the one we call Uranus. Marduk's father, Enki, was originally associated with the Moon; now Marduk gave him the honor of being "number one" planet—the outermost, the one we call Neptune. To hide the forgery and make it appear as though it was always so, the Babylonian version of the Epic of Creation (called *Enuma elish* after its opening words) employed Sumerian terminology for the planetary names, calling the planet NUDIMMUD, "The Artful Creator"— which was exactly what Enki's Egyptian epithet, *Khnum*, had meant.

A celestial counterpart was needed for Marduk's son Nabu. To achieve that, the planet we call Mercury, which was associated with Enlil's young son Ishkur/Adad, was expropriated and allocated to Nabu. Sarpanit, Marduk's spouse to whom he owed his release from the Great Pyramid and the commutation of the sentence of being buried alive in it to that of exile (the first one of the two), was also not forgotten. Settling accounts with Inanna/Ishtar, he deprived her of the celestial association with the planet we call Venus and granted the planet to Sarpanit. (As it happened, while the switch from Adad to Nabu was partly retained in Babylonian astronomy, that of replacing Ishtar by Sarpanit did not take hold.)

Enlil was too omnipotent to be shoved aside. Instead of changing Enlil's celestial position (as the god of the Seventh Planet, Earth) Marduk appropriated to himself the Rank of Fifty that was Enlil's rank, just a rung below Anu's sixty (Enki's numerical rank was forty). That takeover was incorporated into the *Enuma elish* by listing, in the seventh and last tablet of the epic, the Fifty Names of Marduk. Starting with his own name, "Marduk," and ending with his new celestial name, "Nibiru," the list accompanied each name-epithet with a laudatory explanation of its meaning. When the reading of the fifty names during the New Year celebrations was completed, there was no achievement, creative deed, benevolence, lordship, or supremacy left out

... "With the Fifty Names," the last two verses of the epic stated, "the Great Gods proclaimed him; with the title Fifty they made him supreme." An epilogue, added by the priest-ly scribe, made the Fifty Names required reading in Babylon:

> *Let them be kept in mind,*
> *let the lead man explain them;*
> *Let the wise and the knowing*
> *discuss them together;*
> *Let the father recite them*
> *and impart them to his son.*

Marduk's seizure of the supremacy in the heavens was accompanied by a parallel religious change on Earth. The other gods, the Anunnaki leaders—even his direct adver-saries—were neither punished nor eliminated. Rather, they were declared subordinate to Marduk through the gimmick of asserting that their various attributes and powers were transferred to Marduk. If Ninurta was known as the god of husbandry, who had given Mankind agriculture by damming the mountain gushes and digging irrigation canals—the function now belonged to Marduk. If Adad was the god of rains and storms, Marduk was now the "Adad of rains." The list, only partially extant on a Babylonian tablet, began as follows:

Ninurta	= Marduk of the hoe
Nergal	= Marduk of the attack
Zababa	= Marduk of the hand-to-hand combat
Enlil	= Marduk of lordship and counsel
Nabium	= Marduk of numbers and counting
Sin	= Marduk the illuminator of the night
Shamash	= Marduk of justice
Adad	= Marduk of rains

Some scholars have speculated that in this concentration of all divine powers and functions in one hand, Marduk had introduced the concept of one omnipotent god—a step toward the monotheism of the biblical Prophets. But that confuses the belief in one God Almighty with a religion in which one god is just superior to the other gods, a polytheism in which one god dominates the others. In the words of *Enuma elish,* Marduk became "the Enlil of the gods," their "Lord."

No longer residing in Egypt, Marduk/Ra became *Amen,* "The Unseen One." Egyptian hymns to him, nevertheless, proclaimed his supremacy, also connoting the new theology that he was now the "god of gods," "more powerful of might than the other gods." In one set of such hymns, composed in Thebes and discovered written on what is known as the *Leiden Papyrus,* the chapters begin with a description of how after the "islands which are in the midst of the Mediterranean" recognized his name as "high and mighty and powerful," the peoples of "the hill countries came down to thee in wonder; every rebellious country was filled of thy terror." Listing other lands that switched their obedience to Amen-Ra, the sixth chapter continued by describing the god's arrival in the Land of the Gods—as we understand it, Mesopotamia—and the ensuing construction there of Amon's new temple—as we understand it, the Esagil. The text reads almost like that of Gudea's description of all the rare building materials brought over from lands near and far: "The mountains yield blocks of stone for thee, to make the great gates of thy temple; vessels are upon the sea, seafaring craft are at the quays, loaded and navigated unto thy presence." Every land, every people, send propitiatory offerings.

But not only people pay Amen homage; so do all the other gods. Here are some of the verses from the following chapters of the papyrus extolling Amen-Ra as the king of the gods:

The company of the gods which came forth from heaven assembled at thy sight, announcing:
"Great of glory, Lord of Lords . . . He is the Lord!"

The enemies of the Universal Lord are overthrown;
his foes who were in heaven and on Earth are no more.
Thou art triumphant, Amen-Ra!

Thou art the god more powerful of might than all
the other gods. Thou art the sole Sole One.
Universal god:
Stronger than all the cities is thy city Thebes.

Ingeniously, the policy was not to eliminate the other Great Anunnaki but to control and supervise them. When, in time, the Esagila sacred precinct was built with appropriate grandeur, Marduk invited the other leading deities to come and reside in Babylon, in special shrines that were built for each one of them within the precinct. The sixth tablet of the epic in its Babylonian version states that after Marduk's own temple-abode was completed, and shrines for the other Anunnaki were erected, Marduk invited all of them to a banquet. "This is Babylon, the place that is your home!" he said. "Make merry in its precincts, occupy its broad places." By acceding to his invitation, the others would literally have made Babylon what its name—*Bab-ili*—had meant: "Gateway of the gods."

According to this Babylonian version, the other gods took their seats in front of the lofty dais on which Marduk had seated himself. Among them were "the seven gods of destiny." After the banqueting and the performance of all the rites, after verifying "that the norms had been fixed according to all the portents,"

> *Enlil raised the bow, his weapon,*
> *and laid it before the gods.*

Recognizing the symbolic declaration of "peaceful coexistence" by the leader of the Enlilites, Enki spoke up:

May our son, the Avenger, be exalted;
Let his sovereignty be surpassing,
without a rival.

May he shepherd the human race to the end of days;
without forgetting, let them acclaim his ways.

Enumerating all the worshiping duties that the people
were to perform in honor of Marduk and the other gods
gathered in Babylon, Enki had this to say to the other
Anunnaki:

As for us, by his names pronounced,
he is our god!
Let us now proclaim his Fifty Names!

Proclaiming his Fifty Names—granting Marduk the Rank
of Fifty that had been Enlil's and Ninurta's—Marduk became the God of Gods. Not a sole God, but the god to
whom the other gods had to pay obeisance.

If the new religion proclaimed in Babylon was a far cry
from a monotheistic theology, scholars (especially at the
turn of this century) wondered and heatedly debated the
extent to which the notion of a Trinity had originated in
Babylon. It was recognized that Babylon's New Religion
stressed the lineage Enki-Marduk-Nabu and that the divinity
of the Son was obtained from a Holy Father. It was pointed
out that Enki referred to him as "our Son," that his very
name, MAR.DUK, meant "Son of the Pure Place" (P.
Jensen), "Son of the Cosmic Mountain" (B. Meissner),
"Son of the Brilliant Day" (F.J. Delitzsch), "Son of Light"
(A. Deimel) or simply "The True Son" (W. Paulus). The
fact that all those leading Assyriologists were German was
primarily due to the particular interest that the Deutsche
Orient-Gesellschaft—an archaeological society that also
served the political and intelligence-gathering ends of Germany—had conducted an unbroken chain of excavations at
Babylon from 1899 until almost the end of World War I
when Iraq fell to the British in 1917. The unearthing of
ancient Babylon (though the remains were by and large those
from the seventh century B.C.) amid the growing realization
that the biblical creation tales were of Mesopotamian origin,
led to heated scholarly debates under the theme *Babel und*

Bibel—Babylon and Bible, and then to theological ones. Was *Marduk Urtyp Christie?* studies (as one so titled by Witold Paulus) asked, after the tale of Marduk's entombment and subsequent reappearance to become the dominant deity was discovered.

The issue, never resolved, was just let evaporate as post–World War I Europe, and especially Germany, faced more pressing problems. What is certain is that the New Age that Marduk and Babylon ushered in circa 2000 B.C. *manifested itself in a new religion,* a polytheism in which one god dominated all the others.

Reviewing four millennia of Mesopotamian religion, Thorkild Jacobsen (*The Treasures of Darkness*) identified as the main change at the beginning of the second millennium B.C. the emergence of national gods in lieu of the universal gods of the preceding two millennia. The previous plurality of the divine powers, Jacobsen wrote, "required the ability to distinguish, evaluate and choose" not just between the gods but also between good and evil. By assuming all the other gods' powers, Marduk abolished such choices. "The national character of Marduk," Jacobsen wrote (in a study titled *Toward the Image of Tammuz*), created a situation in which "religion and politics became more inextricably linked" and in which the gods, "through signs and omens, actively guided the policies of their countries."

The emergence of guiding politics and religion by "signs and omens" was indeed a major innovation of the New Age. It was not a surprising development in view of the importance that celestial signs and omens had played in determining the true beginning of the zodiacal change and in deciding who would become supreme on Earth. For many millennia it was the word of the Seven Who Determine the Destinies, Anu, Enlil, and the other Anunnaki leaders, who made the decisions affecting the Anunnaki; Enlil, by himself, was the Lord of the Command as far as Mankind was concerned. Now, signs and omens in the heavens guided the decisions.

In the "prophecy texts" (one of which we have earlier quoted) the principal gods played a role alongside or within the framework of the celestial omens. Under the New Age, the celestial omens—planetary conjunctions, eclipses, lunar halos, stellar backgrounds, and so on—were sufficient by themselves, and no godly intervention or participation was required: the heavens alone foretold the fates.

Babylonian texts, and those of neighboring nations in the second and first millennia B.C., are replete with such Omina and their interpretation. A whole science, if one so wishes to call it, developed as time went on, with special *beru* (best translated "fortune teller") priests on hand to interpret observations of celestial phenomena. At first the predictions, continuing the trend that began at the time of the Third Dynasty of Ur, concerned themselves with affairs of state— the fate of the king and his dynasty and the fortunes of the land:

When a halo surrounds the Moon and Jupiter
stands within it, there will be an invasion
of the army of Aharru.

When the Sun reaches its zenith and is dark,
the unrighteousness of the land will come to naught.

When Venus draws near Scorpio, evil winds will
come to the land.

When in the month Siwan Venus shall appear
in Cancer, the king will have no rival.

When a halo surrounds the Sun and its opening
points to the south, a south wind will blow.
If a south wind shall blow on the day of the Moon's
disappearance, it will rain from the heavens.

When Jupiter appears at the beginning of the year,
in that year corn will be plentiful.

The "entrances" of planets into the zodiacal constellations were thought of as particularly important, as signs of the enhancement of the planet's (good or bad) influence. The positions of the planets inside the zodiacal constellations were described by the term *Manzallu* ("stations"), from which the Hebrew plural *Mazzaloth* (II Kings 23:5) comes, and from which a *Mazal* ("luck, fortune") evolved, capable of being good luck or bad luck.

Since not only constellations and planets but also months had been associated with various gods—some, by Babylonian times, adversaries of Marduk—the time of the celestial phenomena grew in importance. One omen, as an example, said: "If the Moon shall be eclipsed in the month Ayaru in the third watch" and certain other planets will be at given positions, "the king of Elam will fall by his own sword . . . his son will not take the throne; the throne of Elam will be unoccupied."

A Babylonian text on a very large tablet (VAT-10564) divided into twelve columns contained instructions for what may or may not be done in certain months: "A king may build a temple or repair a holy place only in Shebat and Adar . . . A person may return to his home in Nissan." The text, called by S. Langdon (*Babylonian Menologies and the Semitic Calendars*) "the great Babylonian Church Calendar," then listed the lucky and unlucky months, even days and even half days, for many personal activities (such as the most favorable time for bringing a new bride into the house).

As the omens, predictions, and instructions increasingly assumed a more personal nature, they verged on the horoscopic. Would a certain person, not necessarily the king, recover from an illness? Will the pregnant mother bear a healthy child? If some times or certain omens were unlucky, how could one ward off the ill luck? In time, incantations were devised for the purpose; one text, for example, actually provided the sayings to be recited to prevent the thinning of a man's beard by appealing to "the star that giveth light" with prescribed utterings. All that was followed by the introduction of amulets in which the warding-off verses were inscribed. In time, too, the material of the amulet (mostly made to be worn on a string around the neck) could also make a difference. If made of hematite, one set of instructions stated, "the man could lose that which he acquired." On the other hand, an amulet made of lapis lazuli assured that "he shall have power."

In the famous library of the Assyrian king Ashurbanipal, archaeologists have found more than two thousand clay tablets with texts pertaining to omens. While the majority dealt with celestial phenomena, not all of them did so. Some dealt with dream-omens, others with the interpretation of "oil and water" signs (the pattern made by oil as it was poured on water), even the significance of animal entrails as they appeared after sacrifices. What used to be astronomy became astrology, and astrology was followed by divinations, fortune-telling, sorcery. R. Camblell Thompson was probably right in titling a major collection of omen texts *The Reports of the Magicians and Astrologers of Nineveh and Babylon*.

Why did the New Age bring all that about? Beatrice Goff (*Symbols of Prehistoric Mesopotamia*) identified the cause as the breakdown of the gods-priests-kings framework that had held society together in the prior millennia. "There was no aristocracy, no priesthood, no intelligentsia" to prevent the situation where "all the affairs of living were inextricably bound up with such 'magical' practices." Astronomy became astrology because, with the Olden Gods gone from their "cult centers," the people were looking at least for signs and omens to guide them in turbulent times.

Indeed, even astronomy itself was no longer what it had been during two millennia of Sumerian achievements. Despite the reputation and high esteem in which "Chaldean" astronomy was held by the Greeks in the second half of the first millennium B.C., it was a sterile astronomy and a far cry from that of Sumer, where so many of the principles, methods, and concepts on which modern astronomy is founded had originated. "There is scarcely another chapter in the history of science where a deep gap exists between the generally accepted description of a period and the results which have slowly emerged from a detailed investigation of the source material," O. Neugebauer wrote in *The Exact Sciences in Antiquity*. "It is evident," he wrote, "that mathematical theory played a major role in Babylonian astronomy as compared with the very modest role of observations." That "mathematical theory," studies of the astronomical tablets of the Babylonians revealed, were column upon column upon column of rows of numbers, imprinted—we use the term purposely—on clay tablets *as though they were computer printouts!* Fig. 161 is a photograph of one such (fragmented) tablet; Fig. 162 is the contents of such a tablet converted to modern numerals.

Not unlike the astronomical codices of the Mayas, that contained page after page after page of glyphs dealing with the planet Venus, but without any indication that they were based on actual Mayan observations but rather followed some data source, so were the Babylonian lists of *predicted* positions of the Sun, Moon, and visible planets extremely detailed and accurate. In the Babylonian instance, however, the position lists (called "Ephemerides") were accompanied by procedure texts on companion tablets in which the rules for computing the ephemerides were given step by step; they contained instructions how, for example, to compute— for over *fifty years in advance*—Moon eclipses by taking into account data from columns dealing with the orbital velocities of the Sun and the Moon and other factors that were needed. *But,* to quote from *Astronomical Cuneiform Texts* by O. Neugebauer, "unfortunately these procedure texts do not contain much of what we would call the 'theory' behind the method."

Yet "such a theory," he pointed out, "must have existed because it is impossible to devise computational schemes of high complication without a very elaborate plan." It was clear from the very neat script and carefully arranged columns and rows, Neugebauer stated, that these Babylonian tablets were *copies* meticulously made from preexisting sources already so neatly and accurately arranged. The mathematics on which the number series were based was the Sumerian sexagesimal system, and the terminology

Figure 161

used—of zodiacal constellations, month names, and more than fifty astronomical terms—was purely Sumerian. There can, therefore, be no doubt that the source of the Babylonian data was Sumerian; all the Babylonians knew was how to use them, by translating into Babylonian the Sumerian "procedure texts."

It was not until the eighth or seventh centuries B.C. that astronomy, in what is called the Neo-Babylonian period, reassumed the observational aspects. These were recorded in what scholars (e.g., A.J. Sachs and H. Hunger, *Astro-*

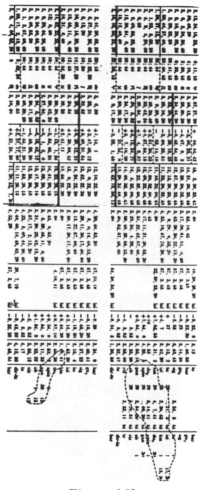

Figure 162

nomical Diaries and Related Texts from Babylonia) call "astronomers' diaries." They believe that Hellenistic, Persian, and Indian astronomy and astrology derived from such records.

The decline and deterioration manifested in astronomy was symptomatic of an overall decline and regression in the sciences, the arts, the laws, the social framework.

One is hard put to find a Babylonian "first," contributed to culture and civilization, that surpassed, or even matched, the countless Sumerian ones. The sexagesimal system and the mathematical theories were retained without improvement. Medicine deteriorated to become little more than sorcery. No wonder that many of the scholars studying the period consider the time when the Old Age of the Sumerian Bull of Heaven gave way to the New Age of the Babylonian Ram a "time of darkness."

The Babylonians, as did the Assyrians and others that followed, retained—almost until the Greek era—the cuneiform script that the Sumerians had devised (based, as we have shown in *Genesis Revisited,* on sophisticated geometric and mathematical theories). But instead of any improvement, the Old Babylonian tablets were written in a more scribbled and less refined script. The many Sumerian references to schools, teachers, homework, were nonexistent in the ensuing centuries. Gone was the Sumerian tradition of literary creativity that bequeathed to future generations, including ours, "wisdom" texts, poetry, proverbs, allegorical tales, and not least of all the "myths" that had provided the data concerning the solar system, the Heavens and Earth, the Anunnaki, the creation of Man. These, it ought to be pointed out, were literary genres that reappear only in the Hebrew Bible about a millennium later. A century and a half of digging up the remains of Babylon produced texts and inscriptions by rulers boasting of military campaigns and conquests, of how many prisoners were taken or heads cut off—whereas Sumerian kings (as, for example, Gudea) boasted in their inscriptions of building temples, digging canals, having beautiful works of art made.

A harshness and a coarseness replaced the former compassion and elegance. The Babylonian king Hammurabi, the sixth in what is called the First Dynasty of Babylon, has been renowned because of his famous legal code, the "Code of Hammurabi." It was, however, just a listing of crimes and their punishments—whereas a thousand years earlier Sumerian kings had promulgated codes of social justice, their laws protecting the widow, the orphan, the weak, and decreeing that "you shall not take away the donkey of a widow," or "you shall not delay the wages of a day laborer.'' Again, the Sumerian concept of laws, intended to direct human conduct rather than punish its faults, reappears only in the biblical Ten Commandments some six centuries after the fall of Sumer. Sumerian rulers cherished the title EN.SI—''Righteous Shepherd.'' The ruler selected by Inanna to reign in Agade (Akkad) whom we call Sargon I in fact bore the name-epithet *Sharru-kin*, ''Righteous King.'' The Babylonian kings (and the Assyrian ones later on) called themselves ''King of the four regions'' and boasted of being ''King of kings'' rather than a ''shepherd'' of the people. (It was greatly symbolic that Judea's greatest king, David, had been a shepherd.)

Missing in the New Age were expressions of tender love. This may sound like an insignificant item in the long list of changes for the worse; but we believe that it was a manifestation of a profound mind-set that went all the way down from the top—from Marduk himself.

The poetry of Sumer included a substantial number of love and lovemaking poems. Some, it is true, were related to Inanna/Ishtar and her relationship with her bridegroom Dumuzi. Others were recited or sung by kings to divine spouses. Yet others were devoted to the common bride and bridegroom, or husband and wife, or parental love and compassion. (Once again, this genre reappears only after many centuries in the Hebrew Bible, in the Book of The Song of Songs.) It seems to us that this omission in Babylonia was not accidental, but part of an overall decline in the role of women and their status as compared to Sumerian times.

The remarkable role of women in all walks of life in Sumer and Akkad and its very marked downgrading upon the rise of Babylon, have been lately reviewed and documented in special studies and several international conferences, such as the ''Invited Lectures on the Middle East at the University of Texas at Austin'' published in 1976 (*The Legacy of Sumer*) by Denise Schmandt-Besserat as editor, and the proceedings at the *33rd Rencontre Assyriologique Internationale* in 1986, whose theme was ''The woman in the ancient Near East.'' The gathered evidence shows that in Sumer and Akkad women engaged not only in household chores like spinning, weaving, milking, or tending to the family and the home, but also were "working profession-als" as doctors, midwives, nurses, governesses, teachers, beauticians, and hairdressers. The textual evidence recently culled from discovered tablets augments the depictions of women in their varied tasks from the earliest recorded times that showed them as singers and musicians, dancers and banquet-masters.

Women were also prominent in business and property management. Records have been found of women managing the family lands and overseeing their cultivation, and then supervising the trade in the resulting products. This was especially true of the "ruling families" of the royal court. Royal wives administered temples and vast estates, royal daughters served not only as priestesses (of which there were three classes) but even as the High Priestess. We have already mentioned Enheduanna, the daughter of Sargon I, who composed a series of memorable hymns to Sumer's great ziggurat-temples. She served as High Priestess at Nan-nar's temple in Ur (Sir Leonard Woolley, who had excavated at Ur, found there a round plaque depicting Enheduanna performing a libation ceremony). We know that the mother of Gudea, Gatumdu, was a High Priestess in the Girsu of Lagash. Throughout Sumerian history, other women held such high positions in the temples and priestly hierarchies. There is no record of a comparable situation in Babylon.

The story of women's role and position in the royal courts was no different. One must refer to Greek sources to find mention of a ruling queen (as distinct from a queen-consort) in Babylonian history—the tale of the legendary Semiramis who, according to Herodotus (I, 184) "held the throne in Babylon" in earlier times. Scholars have been able to establish that she was a historical person, Shammu-ramat. She did reign in Babylon, but only because her husband, the *Assyrian* king Shamshi-Adad, had cap-tured the city in 811 B.C. She served as the royal regent for five years after her husband's death, until their son Adad-Nirari III could assume the throne. "This lady," H.W.F. Saggs wrote in *The Greatness That Was Babylon,* "was obviously very important" because "quite exceptionally for a woman, she is mentioned along with the king in a dedication inscription" (!)

Consort-queens and queen mothers were even more frequent in Sumer; but Sumer could also boast the first-ever queen in her own right, bearing the title LU.GAL ("Great Man") which meant "king." Her name was Ku-Baba; she is recorded in the Sumerian King Lists as "the one who consolidated the foundations of Kish" and headed the Third Dynasty of Kish. There may have been other queens like her during the Sumerian era, but scholars are not certain of their status (i.e. whether they were only queen-consorts or regents for an underage son).

It is noteworthy that even in the most archaic Sumerian depictions in which males were shown naked, females were dressed (Fig. 163a is an example); the exceptions were

Figures 163a, 163b, and 163c

depictions of intercourse, where both were shown naked. As time went on women's dress and garments, as well as their hairdos, became more elaborate and elegant (Fig. 163b, 163c), reflecting their status, education, and noble demeanor. Scholars researching these aspects of the civilizations of the ancient Near East have noted that whereas during the two millennia of Sumerian primacy women were depicted *by themselves* in drawings and in plastic art—hundreds of statues and statuettes that are actual portraits of individual females have been found—there is an almost total absence of such depictions in the post-Sumerian period in the Babylonian empire.

W.G. Lambert titled the paper he had presented at the Rencontre Assyriologique "Goddesses in the Pantheon: A Reflection of Women in Society?" We believe that the situation may well have been the reverse: women's status in society reflected the standing of the goddesses in the pantheon. In the Sumerian pantheon, female Anunnaki played leading roles along with the males from the very beginning. If EN.LIL was "Lord of the Command," his spouse was NIN.LIL, "Lady of the Command"; if EN.KI was "Lord of Earth," his spouse was NIN.KI, "Lady of Earth." When Enki created the Primitive Worker through genetic engineering, Ninharsag was there to be the cocreator. Suffice it to reread the inscriptions of Gudea to realize how many important roles goddesses played in the process that led to the building of the new ziggurat-temple. Suffice it also to point out that one of Marduk's first acts was to transfer to the male Nabu the functions of Nisaba as deity of writing. In fact, all those goddesses that in the Sumerian pantheon held specific knowledge or performed specific functions, were by and large relegated to obscurity in the Babylonian pantheon. When goddesses were mentioned, they were only listed as spouses of the male gods. The same held true for the people under the gods: women were mentioned as wives or daughters, mostly when they were "given" in arranged marriages.

We surmise that the situation reflected Marduk's own bias. Ninharsag, the "Mother of gods and men," was, after all, the mother of his main adversary in the contest for the supremacy on Earth, Ninurta. Inanna/Ishtar was the one who had caused him to be buried alive inside the Great Pyramid. The many goddesses who were in charge of the arts and sciences assisted the construction of the Eninnu in Lagash as a symbol of defiance of Marduk's claims that his time had come. Was there any reason for him to retain the high position and veneration of all these females? Their downgrading in religion and worship was, we believe, reflected in a general downgrading of the status of women in the post-Sumerian society.

An interesting aspect of that was the apparent change in the rules of succession. The source of the conflict between Enki and Enlil was the fact that while Enki was Anu's firstborn, Enlil was the Legitimate Heir because he was born to Anu by a mother who was Anu's half sister. On Earth, Enki repeatedly tried to have a son by Ninharsag, a half sister of his and Enlil's, but she bore him only female offspring. Ninurta was the Legitimate Heir on Earth because it was Ninharsag who bore him to Enlil. Following these rules of succession, it was Isaac who was born to Abraham by his half sister Sarah and not the firstborn Ishmael (the son of the maid Hagar) who became the patriarch's Legitimate Heir. Gilgamesh, king of Erech, was two-thirds (not just one-half) ''divine'' because his mother was a goddess; and other Sumerian kings sought to enhance their status by claiming that a goddess nursed them with mother's milk. All such matriarchal lineages lost their significance when Marduk became supreme. (Maternal lineage became significant again among the Jews at the time of the Second Temple.)

What was the ancient world experiencing at the beginning of the New Age of the 20th century B.C., in the aftermath of international wars, the use of nuclear weapons, the dissolution of a great unifying political and cultural system, the displacement of a boundaryless religion with one of national gods? We at the end of the 20th century A.D. may find it possible to visualize, having ourselves witnessed the aftermath of two world wars, the use of nuclear weapons, the dissolution of a giant political and ideological system, the displacement of centrally controlled and boundless empires by religiously guided nationalism.

The phenomena of millions of war refugees on the one hand, and the rearrangement of the population-map on the other hand, so symptomatic of the events of the twentieth century A.D., *had their counterparts in the twentieth century* B.C.

For the first time there appears in Mesopotamian inscriptions the term *Munnabtutu*, literally meaning "fugitives from a destruction." In light of our twentieth century A.D. experience a better translation would be "displaced persons"—people who, in the words of several scholars, had been "de-tribalized," people who had lost not only their homes, possessions, and livelihoods but also the countries to which they had belonged and were henceforth "stateless refugees," seeking religious asylum and personal safe havens in other peoples' lands.

As Sumer itself lay prostrate and desolate, the remnants of its people (in the words of Hans Baumann, *The Land of Ur*) "spread in all directions; Sumerian doctors and astronomers, architects and sculptors, cutters of seals and scribes, became teachers in other lands."

To all the many Sumerian "firsts," they have thus added one more as Sumer and its civilization came to a bitter end: *the first Diaspora . . .*

Their migrations, it is certain, took them to where earlier groups had gone, such as Harran where Mesopotamia links up with Anatolia, the place to where Terah and his family had migrated and which was already then known as "Ur away from Ur." They undoubtedly stayed (and prospered) there in ensuing centuries, for Abraham sought a bride for Isaac his son among the erstwhile relatives there, and so did Isaac's son Jacob. Their wanderings no doubt also followed in the footsteps of the famed Merchants of Ur, whose loaded caravans and laden ships had blazed trails on land and sea to places near and far. Indeed, one can learn where the "displaced persons" of Sumer went by looking at the foreign cultures that sprouted one after the other in foreign lands—cultures whose script was the cuneiform, whose language included countless Sumerian "loanwords" (especially in the sciences), whose pantheons, even if the gods were called by local names, were the Sumerian pantheon, whose "myths" were the Sumerian "myths," whose tales of heroes (such as of Gilgamesh) were of Sumerian heroes.

How far did the wanderers of Sumer go?

We know that they certainly went to the lands where new nation-states were formed within two or three centuries after the fall of Sumer. While the *Amurru* ("Westerners"), followers of Marduk and Nabu, poured into Mesopotamia and provided the rulers that made up the First Dynasty of Marduk's Babylon, other tribes and nations-to-be engaged in massive population movements that forever changed the Near East, Asia, and Europe. They brought about the emergence of Assyria to Babylon's north, the Hittite kingdom to the northwest, the Hurrian Mitanni to the west, the Indo-Aryan kingdoms that spread from the Caucasus on Babylon's northeast and east, and those of the "Desert peoples" to the south and of the "Sealand people" to the southeast. As we know from the later records of Assyria, Hatti-land, Elam, Babylon and from their treaties with others (in which each one's national gods were invoked), the great gods of Sumer did forgo the "invitation" of Marduk to come and reside within the confines of Babylon's sacred precinct; instead, they mostly became national gods of the new or old-new nations.

It was in such lands that the Sumerian refugees were given asylum all around Mesopotamia, serving at the same time as catalysts for the conversion of their host countries into modern and flourishing states. But some must have ventured to more distant lands, migrating there on their own or, more probably, accompanying the displaced gods themselves.

To the east there stretched the limitless reaches of Asia. Much discussed has been the migratory wave of the Aryans (or Indo-Aryans as some prefer). Originating somewhere southwest of the Caspian Sea, they migrated to what had been the Third Region of Ishtar, the Indus Valley, to repopulate and reinvigorate it. The Vedic tales of gods and heroes that they brought with them were the Sumerian "myths" retold; the notions of Time and its measurement and cycles were of Sumerian origin. It is a safe assumption, we believe, that mingled into the Aryan migration were Sumerian refugees; we say "safe assumption" because Sumerians had to pass that way in order to reach the lands that we call the Far East.

It is generally accepted that within two centuries or so of 2000 B.C. a "mysteriously abrupt change" (in the words of William Watson, *China*) had occurred in China; without any gradual development the land was transformed from one of primitive villages to one with "walled cities whose rulers possessed bronze weapons and chariots and the knowledge of writing." The cause, all agree, was the arrival of migrants from the west—the same "civilizing influences" of Sumer "which can ultimately be traced to the cultural migrations comparable to those which radiated in the West from the Near East"—the migrations in the aftermath of the fall of Sumer.

The "mysteriously abrupt" new civilization blossomed out in China circa 1800 B.C. according to most scholars. The vastness of the country and the sparseness of the earliest evidence offer fertile grounds for scholarly disagreements, but the prevalent opinion is that writing was introduced together with Kingship by the Shang Dynasty; the purpose was significant in itself: to record *omens* on animal bones. The omens were mostly concerned with inquiries for guidance from enigmatic Ancestors.

The writing was monosyllabic and the script was pictographic (from which the familiar Chinese characters evolved into a kind of "cuneiform"—Fig. 164)—both hallmarks of Sumerian writing. Nineteenth-century observations regarding the similarity between the Chinese and Sumerian scripts were the subject of a major study by C.J. Ball (*Chinese and Sumerian*, 1913) that was published under the auspices of Oxford University. It proved conclusively the similarity between the Sumerian pictographs (from which the cuneiform

Figure 164

signs evolved) and the old forms (*Ku Wen*) of Chinese writing.

Ball also tackled the issue of whether this was a similarity stemming only from the expectation that a man or a fish would be drawn pictorially in similar ways even by unrelated cultures. What his research has shown was that not only did the pictographs look the same, but they also (in a material number of instances) were pronounced the same way; this included such key terms as *An* for "heaven" and "god," *En* for "lord" or "chief," *Ki* for "Earth" or "land," *Itu* for "month," *Mul* for "bright/shining" (planet or star). Moreover, when a Sumerian syllable had more than one meaning, the parallel Chinese pictograph had a similar set of varied meanings; Fig. 165 reproduces some of the more than one hundred instances illustrated by Ball.

Recent studies in linguistics, spearheaded by former Soviet scholars, have expanded the Sumerian link to include the whole family of Central and Far Asian or "Sino-Tibetan" languages. Such links form only one aspect of a variety of scientific and "mythological" aspects that recall those of Sumer. The former are especially strong; such aspects as the calendar of twelve months, time counting by dividing the day into twelve double-hours, the adoption of the totally arbitrary device of the zodiac, and the tradition of astronomical observations are entirely of Sumerian origin.

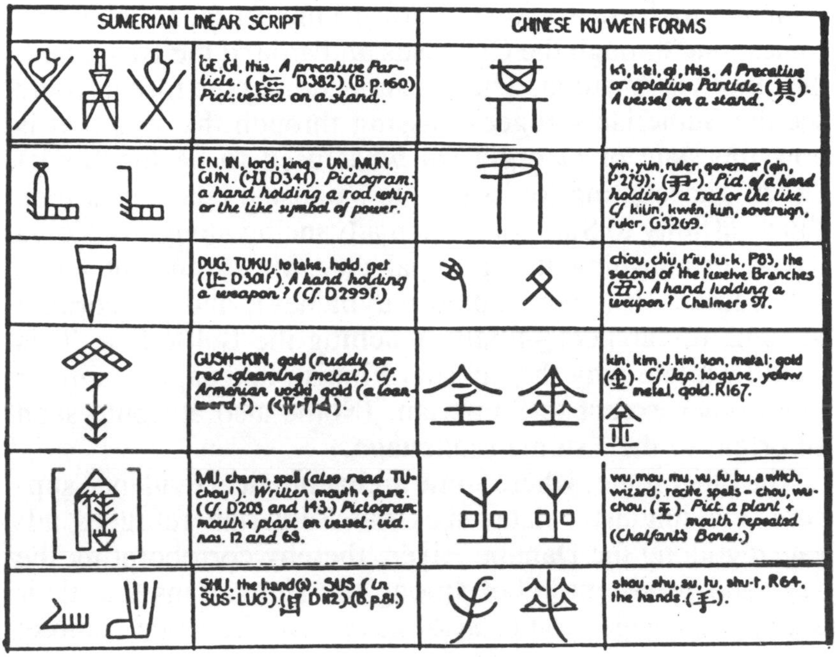

SUMERIAN LINEAR SCRIPT		CHINESE KU WEN FORMS	
	GE, GI. this. A precative Particle. (= D382) (B p.160) Pict: vessel on a stand.		ki, ktí, gi, this. A Precative or optative Particle. (=). A vessel on a stand.
	EN, IN, lord; king = UN, MUN. GUN. (=II D34?). Pictogram: a hand holding a rod, whip, or the like symbol of power.		yin, yin, ruler, governor (qin, P279); (=). Pict. of a hand holding a rod or the like. Cf. kilin, kwún, kun, sovereign, ruler, G3269.
	DUG, TUKU, to take, hold, get (=II- D30I?). A hand holding a weapon? (Cf. D299f.)		chou, chiu, Mu, tu-k, P85, the second of the twelve Branches (=). A hand holding a weapon? Chalmers 97.
	GUSH-KIN, gold (ruddy or red-gleaming metal). Cf. Armenian voski, gold (a loan-word?). (=II=).		kin, kim, J.kin, kon, metal; gold (=). Cf. Jap. kogane, yellow metal, gold. R167.
	MU, charm, spell (also read TU-chou'). Written mouth + pure. (Cf. D205 and 143.) Pictogram: mouth + plant on vessel; vid. nos. 12 and 63.		wu, mou, mu, vu, fu, bu, a witch, wizard; recite spells - chou, wu-chou. (=). Pict. a plant + mouth repeated (Chalfant's Bones.)
	SHU, the hand(s); SUS (in SUS-LUG) (= DII2.)(B p.81)		shou, shu, su, tu, shu-t, R64, the hands. (=).

Figure 165

The "mythological" links are more widespread. Throughout the steppes of Central Asia and all the way from India to China and Japan, the religious beliefs spoke of gods of Heaven and Earth and of a place called *Sumeru* where, at the navel of the Earth, there was a bond that connected the Heaven and the Earth as though the two were two pyramids, one inverted atop the other, linked together as an hourglass with a long and narrow waist. The Japanese *Shintu* religious belief that their emperor is descended of a son of the Sun becomes plausible if one assumes that the reference is not to the star around which Earth orbits, but to the "Sun god" Utu/Shamash; for with the spaceport in the Sinai of which he had been in charge obliterated, and the Landing Place in Lebanon in Mardukian hands, he may well have wandered with bands of his followers to the far reaches of Asia.

As linguistic and other evidence indicates, the Sumerian

Munnabtutu had also gone westward into Europe, using two routes: one through the Caucasus and around the Black Sea, the other via Anatolia. Theories concerning the former route see the Sumerian refugees passing through the area that is now the state of Georgia (in what used to be the Soviet Union), accounting for its people's unusual language which shows affinity to Sumerian, then advancing along the Volga River, establishing its principal city whose ancient name was Samara (it is now called Kuybichev), and—according to some researchers—finally reaching the Baltic Sea. This would explain why the unusual Finnish language is similar to no other except to Sumerian. (Some also attribute such an origin to the Estonian language.)

The other route, where some archaeological evidence supports the linguistic data, sees the Sumerian refugees advancing along the Danube River, thereby corroborating the deep and persistent belief among the Hungarians that their unique language could also have had but one source: Sumerian.

Have Sumerians indeed come this way? The answer might be found in one of the most puzzling relics from antiquity that can be seen where the Danube meets the Black Sea in what was once the Celtic-Roman province of Dacia (now part of Romania). There, at a site called Sarmizegetusa, a series of what researchers have called "calendrical temples" includes what could well be described as *"Stonehenge by the Black Sea."*

Built on several man-made terraces, various structures have been so designed as to form integrated components of a wondrous Time Computer made of stone and wood (Fig. 166). Archaeologists have identified five structures that were in reality rows of round stone "lobes" shaped to form short cylinders, neatly arranged within rectangles formed by sides made of small stones cut to a precise design. The two larger of these rectangular structures contained sixty lobes each, one (the "large old sanctuary") in four rows of fifteen, and the other (the "large new sanctuary") six rows of ten.

Three components of this ancient "calendar city" were round. The smallest is a stone disk made of ten segments

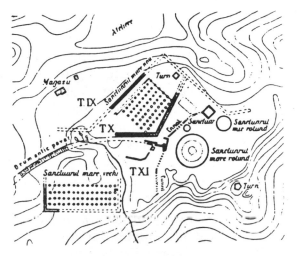

Figure 166

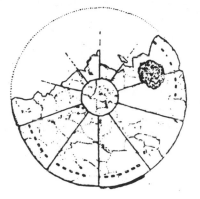

Figure 167

(Fig. 167) in which small stones were embedded to form a circumference—six stones per segment, making a total of sixty. The second round structure, sometimes called the "small round sanctuary," consists of a perfect circle of stones, all precisely and identically shaped, arranged in eleven groups of eight, one of seven and one of six; wider and differently shaped stones, thirteen in number, were placed so as to separate the other grouped stones. There must have been other posts or pillars within the circle, for observation and computing; but none can be determined with certainty. Studies, such as *Il Templo-Calendario Dacico di Sarmizegetusa* by Hadrian Daicoviciu, suggest that this structure served as a lunar-solar calendar enabling a variety of calculations and forecasts, including the proper intercalation between the solar and lunar years by the periodic addition of a thirteenth month. This and the prevalence of the number sixty, the basic number of the Sumerian sexagesimal system, led researchers to discern strong links to ancient Mesopotamia. The similarities, H. Daicoviciu wrote, "could have been neither a coincidence nor an accident." Archaeological and ethnographic studies of the area's history and prehistory in general indicate that at the beginning of the second millennium B.C. a Bronze Age civilization of "nomadic shepherds with a superior social organization" (*Rumania,* an official guidebook) arrived in the area that was until then settled by "a population of simple hand-tillers." The time and the description fit the Sumerian migrants.

The most impressive and intriguing component of this Calendar City is the third round "temple." It consists of two concentric circles surrounding a "horseshoe" in the middle (Fig. 168), bearing an uncanny similarity to Stonehenge in Britain. The outer circle, some 96 feet in diameter,

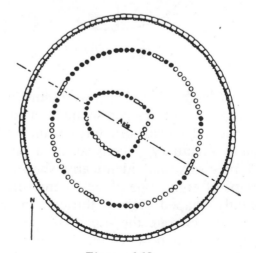

Figure 168

is made up of a ring of 104 dressed andesite blocks that surrounds 180 perfectly shaped oblong andesite blocks, each with a square peg on top as though all were intended to support a movable marker. These uprights are arranged in groups of six; the groups are separated by perfectly shaped horizontal stones, thirty in all. Altogether, then, the outer circle of 104 stones rings an inner circle of 210 (180 + 30) stones.

The second circle, between the outer one and the horseshoe, consists of sixty-eight postholes—akin to the Aubrey holes at Stonehenge—divided into four groups separated by horizontal stone blocks: three each in the northeast and southeast positions and four each in the northwest and southeast positions, giving the "henge" its main northwest–southeast axis and its perpendicular northeast–southwest one. These four grouped markers, one can readily notice, emulate the four Station Stones at Stonehenge.

The final, and immediately obvious, similarity to Stonehenge is the innermost "horseshoe"; it consists of an elliptical arrangement of twenty-one postholes, separated by two horizontal stones on each side from a locking line of thirteen postholes that face to the southeast, leaving no doubt that the principal observational target was the winter solstice Sun. H. Daicoviciu, eliminating some of the wooden posts for simpler visualization, offered a drawing of how the "temple" might have looked (Fig. 169). Noting that the

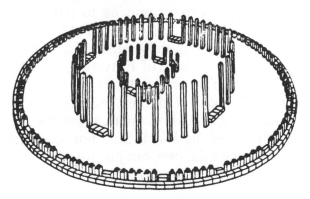

Figure 169

wooden posts were coated with a terra-cotta "plating," Serban Bobancu and other researchers at the National Academy of Romania (*Calendrul de la Sarmizegetusa Regia*) observed that each of those posts "had a massive limestone block as its foundation, a fact that undoubtedly reveals the numerical structure of the sanctuary and proves, as in fact all the others structures do, that the builders wished these structures to last throughout centuries and millennia."

These latest researchers concluded that the "old temple" originally consisted of only fifty-two lobes (4 × 13 rather than the 4 × 15 arrangement) and that there were in effect two calendrical systems geared into one another at Sarmizegetusa: one a solar-lunar calendar with Mesopotamian roots, and another "ritual calendar" geared to fifty-two, akin to the sacred cycle of Mesoamerica and having stellar aspects rather than lunar-solar ones. They concluded that the "stellar era" consisted of four periods of 520 years each (double the 260 of the Mesoamerican Sacred Calendar), and that the ultimate purpose of the calendrical complex was to measure an "era" of 2,080 years (4 × 520)—the approximate length of the Age of Aries.

Who was the mathematical-astronomical genius who had devised all that, and to what purpose?

The spellbinding answer, we believe, also leads to a solution of the enigmas of Quetzalcoatl and the circular observatories that he had built, the god who according to Mesoamerican lore left at one point in time to go back eastward across the seas (promising to return). Was it not just the Enlilite gods who had guided and led the wandering Sumerians, but also Thoth/Ningishzidda (alias Quetzalcoatl), the god of the Game of Fifty-two, who himself had been displaced from his native land?

And was the purpose of all the "stonehenges" in Sumer, and South America, and Mesoamerica, and the British Isles, and on the shores of the Black Sea, not so much to adjust the lunar year to the solar one, not just for calculating Earthly Time but—ultimately—to calculate Celestial Time, the zodiacal Ages?

When the Greeks adopted Thoth as their god Hermes,

they bestowed on him the title *Hermēs Trismegistos,* "Hermes the thrice greatest." Perhaps they recognized that he had thrice guided Mankind in the observation of the beginning of a New Age—the changeover to Taurus, to Aries, to Pisces.

For that was, for those generations of Mankind, when Time began.

ADDITIONAL SOURCES

In addition to sources cited in context, the following periodicals, scholarly studies, and individual works were among the sources consulted:

I. Studies, articles and reports in various issues of the following periodicals and scholarly series:

Abhandlungen für die Kunde des Morgenlandes (Berlin)

Acta Orientalia (Copenhagen and Oslo)

Der Alte Orient (Leipzig)

Alter Orient und Altes Testament (Neukirchen-Vluyn)

American Antiquity (Salt Lake City)

American Journal of Semitic Languages and Literature (Chicago)

American Oriental Series (New Haven)

Analecta Orientalia (Rome)

Anatolian Studies (London)

Annual of the American Schools of Oriental Research (New Haven)

Antigüedades de Mexico (Mexico City)

Archaeology (New York)

Architectura (Munich)

Archiv für Keilschriftforschung (Berlin)

Archiv für Orientforschung (Berlin)

Archiv Orientalni (Prague)

Archives des Sciences physique et naturelles (Paris)

The Assyrian Dictionary (Chicago)

Assyriological Studies (Chicago)

Assyriologische Bibliothek (Leipzig)

Astronomy (Milwaukee)

Babyloniaca (Paris)

Beiträge zur Assyriologie und semitischen Sprachwissenschaft (Leipzig)

Biblica et Orientalia (Rome)

Bibliotheca Mesopotamica (Malibu)

Bibliotheca Orientalis (Leiden)

Biblische Studien (Freiburg)

Bulletin of the American Schools of Oriental Research (Jerusalem and Baghdad)

Centaurus (Copenhagen)

Cuneiform Texts from Babylonian Tablets (London)

Deutsche Akademie der Wissenschaften: Mitteilungen der Institut für Orientforschung (Berlin)

Deutsches Morgenländische Gesellschaft, Abhandlungen (Leipzig)

Ex Oriente Lux (Leipzig)

Grundriss der Theologischen Wissenschaft (Freiburg and Leipzig)

Harvard Semitic Series (Cambridge, Mass.)

Hebrew Union College Annual (Cincinnati)

Icarus (San Diego)

Inca (Lima)

Institut Français d'Archéologie Orientale, Bulletin (Paris)

Iranica Antiqua (Leiden)

Iraq (London)

Isis (London)

Journal of the American Oriental Society (New Haven)

Journal Asiatique (Paris)

Journal of Biblical Literature and Exegesis (Middletown)

Journal of the British Astronomical Association (London)

Journal of Cuneiform Studies (New Haven)

Journal of Egyptian Archaeology (London)
Journal of Jewish Studies (Chichester, Sussex)
Journal of Near Eastern Studies (Chicago)
Journal of the Manchester Egyptian and Oriental Society (Manchester)
Journal of the Royal Asiatic Society (London)
Journal of Semitic Studies (Manchester)
Journal of the Society of Oriental Research (Chicago)
Keilinschriftliche Bibliothek (Berlin)
Klio (Leipzig)
Königliche Gesellschaft der Wissenchaften zu Göttingen: Abhandlungen (Göttingen)
Leipziger semitische Studien (Leipzig)
Mesopotamia: Copenhagen Studies in Assyriology (Copenhagen)
El Mexico Antiguo (Mexico City)
Mitteilungen der altorientalischen Gesellschaft (Leipzig)
Mitteilungen der Deutschen Orient-Gesellschaft (Berlin)
Mitteilungen der vorderasiatisch-aegyptischen Gesellschaft (Berlin)
Mitteilungen des Instituts für Orientforschung (Berlin)
München aegyptologische Studien (Berlin)
Musée du Louvre: Textes Cunéiformes (Paris)
Musée Guimet: Annales (Paris)
The Museum Journal (Philadelphia)
New World Archaeological Foundation: Papers (Provo)
Occasional Papers on the Near East (Malibu)
Oriens Antiquus (Rome)
Oriental Studies (Baltimore)
Orientalia (Rome)
Orientalische Literaturzeitung (Berlin)
Oxford Editions of Cuneiform Inscriptions (Oxford)
Proceedings of the Society of Biblical Archaeology (London)

Publications of the Babylonian Section, University Museum (Philadelphia)

Quellen und Studien zur Geschichte der Mathematik, Astronomie und Physik (Berlin)

Reallexikon der Assyriologie (Berlin)

Recherches d'archéologie, de philosophie et d'histoire (Cairo)

Records of the Past (London)

Revista del Museo Nacional (Lima)

Revista do Instituto Historico e Geografico Brasiliero (Rio de Janeiro)

Revue Archéologique (Paris)

Revue biblique (Paris)

Revue d'Assyriologie et d'archéologie orientale (Paris)

Revue des Etudes Semitique (Paris)

Scientific American (New York)

Service des Antiquites: Annales de l'Egypte (Cairo)

Society of Biblical Archaeology: Transactions (London)

Studi Semitici (Rome)

Studia Orientalia (Helsinki)

Studien zu Bauforschung (Berlin)

Studies in Ancient Oriental Civilizations (Chicago)

Studies in Pre-Columbian Art and Archaeology (Dumbarton Oaks)

Sumer (Baghdad)

Syria (Paris)

Texts from Cuneiform Sources (Locust Valley, N.Y.)

University Museum Bulletin, University of Pennsylvania (Philadelphia)

Vorderasiatische Bibliothek (Leipzig)

Die Welt des Orients (Göttingen)

Wiener Zeitschrift für die Kunde des Morgenlandes (Vienna)

Yale Oriental Series (New Haven)

Zeitschrift der deutschen morgenländischen Gesellschaft (Leipzig)

Zeitschrift für Assyriologie und verwandte Gebiete (Leipzig)

Zeitschrift für die alttestamentliche Wissenschaft (Berlin, Gissen)

Zeitschrift für Keilschriftforschung (Leipzig)

Zenit (Utrecht)

II. Individual Works and Studies:

Abetti, G. *The History of Astronomy.* 1954.

Antoniadi, E.-M. *L'astronomie égyptienne.* 1934.

Armour, R.A. *Gods and Myths of Ancient Egypt.* 1986.

Asher-Greve, J.M. *Frauen in altsumerischer Zeit.* 1985.

Aubier, C. *Astrologie Chinoise.* 1985.

Aveni, A.F. *Skywatchers of Ancient Mexico.* 1980.

————*Empires of Time: Calendars, Clocks and Cultures.* 1989.

Aveni, A.F. (ed.) *Archaeoastronomy in Pre-Columbian America.* 1975.

————*Native American Astronomy.* 1977.

————*Archaeoastronomy in the New World.* 1982.

————*World Archaeoastronomy.* 1989.

Babylonian Talmud

Balfour, M.D. *Stonehenge and its Mysteries.* 1980.

Barklay, E. *Stonehenge and its Earthworks.* 1895.

Barrois, A.-G. *Manuel d'Archéologie Biblique.* 1939.

Barton, G.A. *The Royal Inscriptions of Sumer and Akkad.* 1929.

Benzinger, I. *Hebräische Archäologie.* 1927.

Bittel, K. (ed.) *Anatolian Studies Presented to Hans Gustav Güterbock.* 1974.

Bobula, I. *Sumerian Affiliations.* 1951.

————*The Origin of the Hungarian Nation.* 1966.

Boissier, A. *Choix de Textes.* 1905–6.

Boll, F. and Bezold, C. *Sternglaube und Sternbedeutung.* 1926.

Boll, F., Bezold, C. and Gundel, W. *Sternglaube, Sternreligion und Sternorakel.* 1927.

Bolton, L. *Time Measurement.* 1924.

Borcchardt, L. *Beiträge zur Ägyptische Bauforschung und Altertumskunde.* 1937–1950.

Bottero, J. and Kramer, S.N. *Lorsque les dieux faisaient l'Homme.* 1989.

Brown, P.L. *Megaliths, Myths and Men.* 1976.

Brugsch, H.K. *Nouvelle Recherches sur la Division de l'Anneé des Anciens Égyptiens.* 1856.

———*Thesaurus Inscriptionum Aegyptiacarum.* 1883.

———*Religion und Mythologie der alten Aegypter.* 1891.

Budge, E.A.W. *The Gods of the Egyptians.* 1904.

Burl, A. *The Stone Circles of the British Isles.* 1976.

———*Prehistoric Avebury.* 1979.

Canby, C.A. *A Guide to the Archaeological Sites of the British Isles.* 1988.

Caso, A. *Calendario y Escritura de las Antiguas Culturas de Monte Alban.* 1947.

———*Los Calendarios Prehispanicos.* 1967.

Charles, R.H. *The Apocrypha and Pseudoepigrapha of the Old Testament.* 1976 edition.

Chassinat, E.G. *Le Temple de Dendera.* 1934.

Chiera, E. *Sumerian Religious Texts.* 1924.

Childe, V.G. *The Dawn of European Civilization.* 1957.

Chippindale, C. *Stonehenge Complete.* 1983.

Clay, A.T. *Babylonian Records in the Library of J. Pierpont Morgan.* 1912–1923.

Cornell, J. *The First Stargazers.* 1981.

Cottrell, A. (ed.) *The Encyclopedia of Ancient Civilizations.* 1980.

Craig, J.A. *Astrological-Astronomical Texts in the British Museum.* 1899.

Dalley, S. *Myths from Mesopotamia*. 1989.

Dames, M. *The Silbury Treasure*. 1976.

—— *The Avebury Cycle*. 1977.

Daniel, G. *The Megalithic Builders of Western Europe*. 1962.

Dhorme, P. *La Religion Assyro-babylonienne*. 1910.

Dubelaar, C.N. *The Petroglyphs in the Guianas and Ancient Areas of Brazil and Venezuela*. 1986.

Dumas, F. *Dendera et le temple d'Hathor*. 1969.

Dunand, M. *Fouilles de Byblos*. 1939–1954.

Durand, J.-M. (ed.) *La femme dans le Proche-Orient antique*. 1986.

Eichhorn, W. *Chinese Civilization*. 1980.

Eichler, B.L. (ed.) *Kramer Anniversary Volume*. 1976.

Eisler, R. *Weltenmantel und Himmelszeit*. 1910.

—— *The Royal Art of Astronomy*. 1946.

Emery, W.B. *Archaic Egypt*. 1961.

Endrey, A. *Sons of Nimrod*. 1975.

Epping, J. *Astronomisches aus Babylon*. 1889.

Falkenstein, A. *Archaische Texte aus Uruk*. 1936.

—— *Sumerische Götterlieder*. 1959.

Falkenstein, A. and von Soden, W. *Sumerische und Akkadische Hymnen und Gebete*. 1953.

Fischer, H.G. *Dendera in the Third Millenium B.C.* 1968.

Flornoy, B. *Amazone—Terres et Homme*. 1969.

Fowles, J. and Brukoff, B. *The Enigma of Stonehenge*. 1980.

Frankfort, H. *The Problem of Similarity in Ancient Near Eastern Religions*. 1951.

—— *The Art and Architecture of the Ancient Orient*. 1969.

Gaster, T.H. *Myth, Legend and Custom in the Old Testament*. 1969.

Gauquelin, M. *The Scientific Basis of Astrology*. 1969.

Gibson, Mc. and Biggs, R.D. (eds.) *Seals and Sealing in the Ancient Near East*. 1977.

Gimbutas, M. *The Prehistory of Eastern Europe*. 1956.

Girshman, R. *L'Iran et la migration des Indo-aryens et des iraniens*. 1977.

Grayson, A.K. *Assyrian and Babylonian Chronicles*. 1975.

——— *Babylonian Historical Literary Texts*. 1975.

Gressmann, H. (ed.) *Altorientalische Texte zum alten Testament*. 1926.

Grimm, J. *Teutonic Mythology*. 1900.

Haddingham, E. *Early Man and the Cosmos*. 1984.

Hallo, W.W. and Simpson, W.K. *The Ancient Near East: A History*. 1971.

Hartmann, J. (ed.) *Astronomie*. 1921.

Heggie, D.C. *Megalithic Science*. 1981.

Heggie, D.C. (ed.) *Archaeoastronomy in the Old World*. 1982.

Higgins, R. *Minoan and Mycenaean Art*. 1967.

Hilprecht, H.V. *Old Babylonian Inscriptions*. 1896.

Hilprecht Anniversary Volume. 1909.

Hodson, F.R. (ed.) *The Place of Astronomy in the Ancient World*. 1974.

Holman, J.B. *The Zodiac: The Constellations and the Heavens*. 1924.

Hommel, F. *Die Astronomie der alten Chaldäer*. 1891.

——— *Aufsätze und Abhandlungen*. 1892–1901.

Hooke, S.H. *Myth and Ritual*. 1933.

——— *The Origins of Early Semitic Ritual*. 1935.

——— *Babylonian and Assyrian Religion*. 1962.

Hoppe. E. *Mathematik und Astronomie im Klassichen Altertums*. 1911.

Ibarra Grasso, D.E. *Ciencia Astronomica y Sociologia*. 1984.

Jastrow, M. *Die Religion Babyloniens und Assyriens*. 1905–1912.

Jean, C.-F. *La religion sumerienne.* 1931.

Jensen, P. *Die Kosmologie der Babylonier.* 1890.

———— *Texte zur assyrisch-babylonischen Religion.* 1915.

Jeremias, A. *Das alter der babylonischen Astronomie.* 1908.

Joussaume, R. *Dolmens for the Dead.* 1988.

Kees, H. *Der Götterglaube im Alten Aegypten.* 1941.

Keightly, D. *Sources of Shang History.* 1978.

Keightly, D. (ed.) *The Origins of Chinese Civilization.* 1983.

Kelly-Buccellati, M. (ed.) *Studies in Honor of Edith Porada.* 1986.

King, L.W. *Babylonian Magic and Sorcery.* 1896.

———— *Babylonian Religion and Mythology.* 1899.

———— *Cuneiform Texts from Babylonian Tablets.* 1912.

Koldewey, R. *The Excavations at Babylon.* 1914.

Komoroczy, G. *Sumer es Magyar?* 1976.

Kramer, S.N. *Sumerian Mythology.* 1961

———— *The Sacred Marriage Rite.* 1980.

———— *In the World of Sumer.* 1986.

Kramer, S.N. and Maier, J. (eds.) *Myths of Enki, the Crafty God.* 1989.

Krickberg, W. *Felsplastik und Felsbilder bei den Kulturvolkern Altameriker.* 1969.

Krupp, E.C. *Echoes of Ancient Skies: The Astronomies of Lost Civilizations.* 1983.

Krupp, E.C. (ed.) *In Search of Ancient Astronomies.* 1978.

———— *Archaeoastronomy and the Roots of Science.* 1983.

Kugler, F.X. *Die babylonische Mondrechnung.* 1900.

———— *Sternkunde und Sterndienst in Babylon.* 1907–1913.

———— *Im Bannkreis Babels.* 1910.

———— *Alter und Bedeutung der babylonischen Astronomie und Astrallehre,* 1914.

Lambert, B.W.L. *Babylonian Wisdom Literature.* 1960.

Langdon, S. *Sumerian and Babylonian Psalms*. 1909.

—— *Tablets from the Archives of Drehem*. 1911.

—— *Die neubabylonischen Koenigs inschriften*. 1912.

—— *Babylonian Wisdom*. 1923.

—— *Babylonian Penitential Psalms*. 1927.

Langdon, S. (ed.) *Oxford Editions of Cuneiform Texts*. 1923.

Lange, K. and Hirmer, M. *Egypt: Architecture, Sculpture, Painting*. 1968.

Lathrap, D.W. *The Upper Amazon*. 1970.

Lehmann, W. *Einige probleme centralamerikanische kalenders*. 1912.

Leichty, E., Ellis, M. de J. and Gerardi, P. (eds.) *A Scientific Humanist: Studies in Memory of Abraham Sachs*. 1988.

Lenzen, H.J. *Die entwicklung der Zikkurat*. 1942.

Lesko, B.S. (ed.) *Women's Earliest Records from Ancient Egypt and Western Asia*. 1989.

Lidzbarski, M. *Ephemeris für Semitische Epigraphik*. 1902.

Luckenbill, D.D. *Ancient Records of Assyria and Babylonia*. 1926–7.

Ludendorff, H. *Über die Entstehung der Tzolkin-Periode im Kalender der Maya*. 1930.

—— *Das Mondalter in der Inschriften des Maya*. 1931.

Lutz, H.F. *Sumerian Temple Records of the Late Ur Dynasty*. 1912.

Mahler, E. *Biblische Chronologie*. 1887.

—— *Handbuch der jüdischen Chronologie*. 1916.

Maspero, H. *L'Astronomie dans la Chine ancienne*. 1950.

Menon, C.P.S. *Early Astronomy and Cosmology*. 1932.

Mosley, M. *The Maritime Foundations of Andean Civilization*. 1975.

Needham, J. *Science and Civilization in China*. 1959.

Neugebauer, O. *Astronomical Cuneiform Texts*. 1955.

——— *A History of Ancient Mathematical Astronomy.* 1975.

Neugebauer, P.V. *Astronomische Chronologie.* 1929.

Newham, C.A. *The Astronomical Significance of Stonehenge.* 1972.

Niel, F. *Stonehenge—Le Temple mystérieux de la préhistoire.* 1974.

Nissen, H.J. *Grundzüge einer Geschichte der Frühzeit des Vorderen Orients.* 1983.

Oates, J. *Babylon.* 1979.

O'Neil, W.M. *Time and the Calendars.* 1975.

Oppenheim, A.L. *Ancient Mesopotamia* (1964; revised 1977).

Pardo, L.A. *Historia y Arqueologia del Cuzco.* 1957.

Parrot, A. *Tello.* 1948.

——— *Ziggurats et Tour de Babel.* 1949.

Petrie, W.M.F. *Stonehenge: Plans, Description and Theories.* 1880.

Piggot, S. *Ancient Europe.* 1966.

Ponce-Sanguines, C. *Tiwanaku: Espacio, Tiempo y Cultura.* 1977.

Porada, E. *Mesopotamian Art in Cylinder Seals.* 1947.

Pritchard, J.B. (ed.) *Ancient Near Eastern Texts Relating to the Old Testament.* 1969.

Proceedings of the 18th Rencontre Assyriologique Internationale. 1972.

Radau, H. *Early Babylonian History.* 1900.

Rawlinson, H.C. *The Cuneiform Inscriptions of Western Asia.* 1861–84.

Rawson, J. *Ancient China.* 1980.

Rice, C. *La Civilizacion Preincaica y el Problema Sumerologico.* 1926.

Rivet, P. *Los origines del hombre americano.* 1943.

Rochberg-Halton, F. (ed.) *Language, Literature and History.* 1987.

Roeder, G. *Altaegyptische Erzählungen und Märchen.* 1927.

Rolleston, F. *Mazzaroth, or the Constellations.* 1875.

Ruggles, C.L.N. *Megalithic Astronomy.* 1984.

Ruggles, C.L.N. (ed.) *Records in Stone.* 1988.

Ruggles, C.L.N. and Whittle, A.W.R. (eds.) *Astronomy and Society in Britain During the Period 4000–1500 B.C.* 1981.

Sasson, J.M. (ed.) *Studies in Literature from the Ancient Near East Dedicated to Samuel Noah Kramer.* 1984.

Saussure, L. de *Les Origines de l'Astronomie Chinoise.* 1930.

Sayce, A.H. *Astronomy and Astrology of the Babylonians.* 1874.

——— *The Religion of the Babylonians.* 1888.

Schiaparelli, G. *L'Astronomia nell'Antico Testamento.* 1903.

Schwabe, J. *Archetyp und Tierkreis.* 1951.

Sertima, I.V. *They Came Before Columbus.* 1976.

Shamasashtry, R. *The Vedic Calendar.* 1979.

Sivapriyananda, S. *Astrology and Religion in Indian Art.* 1990.

Sjöberg, A.W. and Bergmann, E. *The Collection of Sumerian Temple Hymns.* 1969.

Slosman, A. *Le zodiaque de Denderah.* 1980.

Smith, G.E. *Ships as Evidence of the Migrations of Early Cultures.* 1917.

Spinden, H.J. *Origin of Civilizations in Central America and Mexico.* 1933.

Sprockhoff, E. *Die nordische Megalitkultur.* 1938.

Starr, I. *The Rituals of the Diviner.* 1983.

Steward, J.H. (ed.) *Handbook of South American Indians.* 1946.

Stobart, C. *The Glory That Was Greece.* 1964.

Stoepel, K.T. *Südamerikanische Prähistorische Tempel und Gottheiten*. 1912.

Stücken, E. *Beiträge zur orientalischen Mythologie*. 1902.

The Sumerian Dictionary of the University Museum, University of Pennsylvania. 1984–

Tadmor, H. and Weinfeld, M. (eds.) *History, Historiography and Interpretation*. 1983.

Talmon, Sh. *King, Cult and Calendar in Ancient Israel*. 1986.

Taylor, L.W. *The Mycenaeans*. 1966.

Tello, J.C. *Origen y Desarrollo de las Civilizaciones Prehistoricas Andinas*. 1942.

Temple, J.E. *Maya Astronomy*. 1930.

Thom, A. *Megalithic Sites in Britain*. 1967.

Thomas, D.W. (ed.) *Documents from Old Testament Times*. 1961.

Thompson, J.E.S. *Maya History and Religion*. 1970.

Trimborn, H. *Die Indianischen Hochkulturen des Alten Amerika*. 1963.

Van Buren, E.D. *Clay Figurines of Babylonia and Assyria*. 1930.

——— *Religious Rites and a Ritual in the Time of Uruk IV–III*. 1938.

Vandier, J. *Manuel d'Archéologie Égyptienne*. 1952–58.

Virolleaud, Ch. *L'Astronomie Chaldéenne*. 1903–8.

Ward, W.A. *Essays on the Feminine Titles of the Middle Kingdom*. 1986.

Weidner, E.F. *Alter und Bedeutung der babylonischen Astronomie und Astrallehre*. 1914.

——— *Handbuch der babylonischen Astronomie*. 1915.

Wiener, L. *Africa and the Discovery of America*. 1920.

——— *Mayan and Mexican Origins*. 1926.

Wilford, J.N. *The Mapmakers*. 1982.

Williamson, R.A. (ed.) *Archaeoastronomy in the Americas*. 1978.

Winckler, H. *Himmels- und Weltenbilder der Babylonier.* 1901.

Wolkstein, D. and Kramer, S.N. *Inanna, Queen of Heaven and Earth.* 1983.

Wuthenau, A. von *Unexpected Faces in Ancient America.* 1980.

Ziolkowsky, M.S. and Sadowski, R.M. (eds.) *Time and Calendars in the Inca Empire.* 1989.

INDEX